Topics in Ordinary Differential Equations

William D. Lakin
Old Dominion University, Norfolk, Virginia

David A. Sanchez
The University of New Mexico, Albuquerque

DOVER PUBLICATIONS, INC.
NEW YORK

To Jane and Joan

Published in Canada by General Publishing Company, Ltd., 30 Lesmill Road, Don Mills, Toronto, Ontario.
Published in the United Kingdom by Constable and Company, Ltd., 10 Orange Street, London WC2H 7EG.

This Dover edition, first published in 1982, is an unabridged and slightly corrected republication of the work originally published by Prindle, Weber & Schmidt, Inc., Boston, in 1970 under the title *Topics in Ordinary Differential Equations: A Potpourri*.

Manufactured in the United States of America
Dover Publications, Inc., 180 Varick Street, New York, N.Y. 10014

Library of Congress Cataloging in Publication Data

Lakin, William D.
    Topics in ordinary differential equations.

    "Unabridged and slightly corrected republication of the work originally published by Prindle, Weber & Schmidt, Inc., Boston, in 1970"—T.p. verso.
    Includes bibliographical references and index.
    1. Differential equations. I. Sánchez, David A. II. Title.
QA372.L23   1982   515.3'52   82-7258
ISBN 0-486-61606-1   AACR2

Preface

The idea for this book grew out of a series of informal discussions over coffee during the fall of 1968 at UCLA. We were concerned that many students interested in applied mathematics were formally exposed to the subject only at the graduate level, by which time many were firmly committed to other areas in mathematics or the natural sciences. This lack of exposure is especially evident in the area of ordinary differential equations, where the usual undergraduate course barely touches (at best) some of the interesting ideas and techniques by which one obtains analytic approximations to solutions.

A partial solution to the problem seemed to be the introduction of more courses in applied mathematics at the undergraduate level, and this book was developed with such courses in mind. We assume that the student has taken a basic introductory course in ordinary differential equations. With the exception of one example, however, no knowledge of complex variables is needed, though an understanding of the basic properties of complex numbers would be helpful to the reader.

Although we visualized a course in ordinary differential equations, we have tried to make this book as flexible as possible. Thus, it might be suitable for a portion of a methods of applied mathematics sequence, an undergraduate seminar, or a mathematics course for scientists and engineers. However, our main goal was to get the student interested in applied mathematics and ordinary differential equations, not to supplement already existing courses.

The thread that connects the various topics in this book is perturbation methods, regular and singular. Chapter 1 deals with asymptotic expansions, and both coordinate and parameter expansions are introduced. In Chapter 2 some methods of obtaining asymptotic approximations to the solutions of linear second-order ordinary differential equations containing a large parameter are discussed. Both a straightforward method and the more versatile comparison equation method are employed. Error bounds, the construction of higher approximations and the connection problem are also considered.

In Chapter 3 some simple singular perturbation problems are solved using the method of matched asymptotic expansions. To illustrate the basic ideas and give the students a feeling for the method and for the location of the boundary layers, a model problem is used whose exact solution is known.

Chapter 4 discusses the existence of periodic solutions, first for linear second-order equations, then for nonlinear equations. In most cases a small parameter is present and the object of the chapter is to give the reader some "nonlinear horse sense," then to give him several techniques by which he can back up his intuition with approximate solutions.

Finally in Chapter 5 the relation between the Sturm-Liouville problem and the calculus of variations is studied with the intent of obtaining approximate solutions and eigenvalues. A brief introduction to the "simplest" problem of the calculus of variations is given, followed by a study of the Ritz method. The method of Galerkin, applied both to boundary value problems and (briefly) to the approximation of periodic solutions, is also discussed.

The presentation is somewhat informal since the authors felt this would make it easier for the reader to gain a basic understanding of some of the ideas and difficulties involved. The problems are intended to further strengthen this understanding, and the references at the end of each chapter will provide more background as well as related topics. This is in no way a definitive text and for the reader we hope it is just the beginning.

William Lakin is responsible for the material in Chapters 1, 2 and 3, and David Sanchez for that in Chapters 4 and 5. This material was developed in courses each taught at UCLA in 1968–1969. The authors wish to thank those undergraduate students who contributed directly or indirectly to the writing of this book. They also wish to thank Mrs. Ruth Goldstein for her excellent typing and preparation of the manuscript. Finally, they extend their thanks to Professor W. A. Harris, Jr. for his critical comments and review.

William D. Lakin
David A. Sanchez

Contents

1

Asymptotic Expansions

1.1 Introduction

In this chapter, some of the basic ideas of the theory of asymptotic expansions will be introduced. First, however, some motivation for considering asymptotic expansions will be presented.

In most calculus courses, a question considered is: Given a real valued function $f(x)$ of a real variable x containing a number x_0 in its domain of definition D, is there a power series expansion of the form

$$\sum_{j=0}^{\infty} a_j(x - x_0)^j \tag{1}$$

with a nonzero radius of convergence which provides a valid representation for f on I, the interval of convergence of the power series? If $f(x)$ has uniformly bounded derivatives of all orders at each point in I, the answer to this question is yes. Further, the power series is uniquely determined and

$$a_j = \frac{f^{(j)}(x_0)}{j!}, \tag{2}$$

where $f^{(j)}(x_0)$ denotes the jth derivative of $f(x)$ evaluated at x_0. In this case, the power series expansion (1) is called the *Taylor series* of the function $f(x)$ about the point x_0, and is uniquely determined. Also, if $x_0 < x_1$, the closed interval $[x_0, x_1]$ is in the domain of f, and $f^{(n+1)}(x)$ exists for all $x \in [x_0, x_1]$, then Taylor's theorem states that there is an $\tilde{x} \in (x_0, x_1)$ such that

$$f(x_1) = \sum_{j=0}^{n} \frac{f^{(j)}(x_0)}{j!} (x_1 - x_0)^j + R_n, \qquad (3)$$

where

$$R_n = \frac{f^{(n+1)}(\tilde{x})}{(n+1)!} (x_1 - x_0)^{n+1}. \qquad (4)$$

Let B be the uniform bound for $f^{(n+1)}(x)$ on (x_0, x_1), i.e.

$$|f^{(n+1)}(x)| \leq B \qquad \text{for all} \qquad x \in (x_0, x_1).$$

Then, from (4)

$$|R_n| \leq \frac{B}{(n+1)!} (x_1 - x_0)^{n+1},$$

and thus the error introduced by using only n terms of the Taylor series for $f(x_1)$ is of the same order of magnitude as the first term in the series which is neglected.

Unfortunately, the theoretical and computational use of a Taylor series representation often poses serious problems. Suppose, for example, that we are given the Taylor series for $f(x)$ about x_0 and a point $a > x_0$ in both the domain D of f and the interval of convergence I of the power series. Consider the computational problem:

> Using relation (3), compute $f(a)$ where n in (3) is an integer such that $|R_n| \leq \epsilon \ll 1$.

The constant $\epsilon > 0$ gives a bound on the allowable error. At the kth stage of the numerical procedure, we must thus compute a bound for $|R_k|$. If $|R_k| \leq \epsilon$, we can take $n = k$ in (3) and hence

$$\left| f(a) - \sum_{j=0}^{k} \frac{f^{(j)}(x_0)}{j!} (a - x_0)^j \right| \leq \epsilon. \qquad (5)$$

On the other hand, if $|R_k| > \epsilon$, we must compute at least one additional term in the Taylor series.

If $a - x_0$ is large, it may be necessary to compute a large number of terms in the Taylor series before satisfying condition (5). This may not be practical, even with the aid of a modern high-speed computer. In such cases, it is natural to seek a different representation for $f(x)$ which makes the computational problem more manageable.

Often, such an alternate representation takes the form of an *asymptotic expansion*

$$\sum_{n=0}^{\infty} a_n g_n(x),$$

where the functions $g_n(x)$ are determined by the nature of the computational problem.

With appropriate conditions on f, the Taylor series for $f(x)$ on $I \cap D$ is an exact representation in the sense that if we were to sum the entire Taylor series with $x = a$, we would obtain $f(a)$ exactly. Asymptotic expansions, however, may be, and usually are, divergent. They can thus provide only an approximate representation for f. However, as the amount of work needed to obtain $f(a)$ from the Taylor series increases, it is usually the case that the asymptotic expansion becomes a better and better approximation to $f(a)$. The divergence of the series does not cause practical difficulties since in most applications, only the first few terms are needed to achieve the desired accuracy.

Suppose we wish to find the solution of a differential equation having variable coefficients. If the coefficients are analytic on some real interval $|x - x_0| < r_0$, a basic theorem states that we may obtain a solution of the differential equation in the form of a power series expansion which converges in $|x - x_0| < r_0$. However, if the coefficients are not analytic in the interval we wish to consider, it is natural to ask if the solution can still be represented by some expansion. The answer is often yes, with the representation being asymptotic.

As will be seen, a convergent power series expansion is also an asymptotic expansion. However, the preceding discussion indicates that asymptotic expansions are useful entities in their own right which complement the notion of convergent power series expansions.

1.2 Order Operators and Asymptotic Sequences

Suppose you are asked the question "How far have you walked today?" Your answer, depending on how active a person you are or the time of day, might be a few feet, less than a mile, a couple of miles, or about ten miles. These are all satisfactory answers since your questioner does not expect an exact answer such as two miles, four feet, and seven inches. In mathematics, the same sort of situation often arises. We might want to know how large the error is in some approximation or how large a function is on a certain interval. An exact answer would be nice, but might be unobtainable, and we would be satisfied with having a bound or order of magnitude estimate. Order operators allow us to give such estimates.

Let $f(x)$ and $g(x)$ be two real valued functions of a real variable defined on an open interval I, e.g., x is in I if

$$a < x < b.$$

The constants a and b might be $+\infty$ or $-\infty$, respectively. Let x_0 be a limit point of I. Then, by a neighborhood U of x_0, we will mean the open interval

$$|x - x_0| < \delta \tag{6}$$

if x_0 is finite; we will mean the set of all x such that

$$x > \delta \tag{7}$$

if $x_0 = +\infty$; or the set of all x such that

$$x < -\delta \tag{8}$$

if $x_0 = -\infty$. In (6), δ is a positive constant, usually considered to be small. In (7) and (8), δ is an arbitrarily large positive number. The order operators O ("big O") and o ("little o") are defined as follows:

(i) $f(x)$ is "*big O*" or order of $g(x)$, and we write $f(x) = O(g(x))$, if there exists a positive constant C such that $|f(x)| \leq C|g(x)|$ for all x in I.

(ii) $f(x)$ is "*big O*" or order of $g(x)$ *as x goes to x_0*, and we write $f(x) = O(g(x))$ as $x \to x_0$, if there exists a positive constant C and a neighborhood U of x_0 such that $|f(x)| \leq C|g(x)|$ for all x in $U \cap I$.

(iii) $f(x)$ is "*little o*" of $g(x)$ *as x goes to x_0*, and we write $f(x) = o(g(x))$ as $x \to x_0$, if given any $\mu > 0$ there exists a neighborhood U of x_0 such that $|f(x)| < \mu|g(x)|$ for all x in $U \cap I$.

The restriction of x to $U \cap I$ in (ii) and (iii) is necessary as x_0 could be an endpoint of I. In this case, $f(x)$ and $g(x)$ would not be defined for all x in U.

Suppose that $g(x)$ does not vanish for any x in I. Then, $f = O(g)$ simply means that f/g is bounded. Similarly, $f = o(g)$ means that f/g tends to zero as $x \to x_0$. It is also clear from (ii) and (iii) that if $f = o(g)$ as $x \to x_0$, then $f = O(g)$ as $x \to x_0$. However, $f = O(g)$ as $x \to x_0$ does not imply $f = o(g)$ as $x \to x_0$. It should also be noted that if $f(x) = O(g(x))$ and $g(x) = O(h(x))$ then $f(x) = O(h(x))$. Similarly,

$$f(x) + g(x) = O(h(x)).$$

Both types of order operators suppress information since they indicate only orders of magnitude. For example, writing $f = O(g)$ tells us only that a finite bound exists for f/g but gives no indication of its size. Writing $f = o(g)$ suppresses even more information since it only says that f/g goes to zero as $x \to x_0$ but does not say anything about how fast it tends to zero. The information given by the order operators is thus quite vague. However, it is exactly this vagueness which allows us to employ them in a great variety of situations.

Examples

$$\sin x = O(1) \qquad\qquad -\infty < x < +\infty$$

$$x^2 = O(x) \qquad\qquad |x| < 1$$

$$e^x - 1 = O(x) \qquad\qquad |x| < 1$$

$$x = O(x^2) \qquad\qquad x \to \infty$$

$$(\log x)^{-1} = O(1) \qquad\qquad x \to \infty$$

$$(x - x_0)^2 = o(x - x_0) \qquad x \to x_0 \quad (x_0 \text{ finite})$$

$$\frac{1}{x^2} = o\left(\frac{1}{x}\right) \qquad\qquad x \to \infty$$

$$e^{-x} = o(x^{-n}) \qquad\qquad x \to \infty \quad (n \text{ arbitrary})$$

Suppose that $\{g_n(x)\}$ is an ordered sequence of functions. This sequence of functions is called an *asymptotic sequence* as $x \to x_0$ if there is a neighborhood U of x_0 on which none of the $g_n(x)$ vanish if $x \neq x_0$ and if, for $n = 0, 1, 2, \ldots,$

$$g_{n+1}(x) = o\big(g_n(x)\big) \quad \text{as} \quad x \to x_0 .$$

For example, if x_0 is finite, $\{(x - x_0)^n\}$ is an asymptotic sequence as $x \to x_0$. The sequence $\{x^{-n}\}$ is an asymptotic sequence as $x \to \infty$.

Using the basic definition above, we may obtain slightly more general asymptotic sequences. Let $h(x)$ be a function defined on U and suppose that $h(x)$ does not vanish on U for $x \neq x_0$. Let $\{g_n(x)\}$ be an asymptotic sequence as $x \to x_0$. Define the sequence $\{h_n(x)\}$ by

$$h_n(x) = h(x)g_n(x).$$

Then, $\{h_n(x)\}$ is also an asymptotic sequence as $x \to x_0$.

1.3 Asymptotic Expansions with Respect to Coordinates

Given a function $f(x)$ and an asymptotic sequence $\{g_n(x)\}$ as $x \to x_0$, the formal series

$$\sum_{n=0}^{\infty} a_n g_n(x) \quad (a_n = \text{constant}), \tag{9}$$

not necessarily convergent, is said to be an *asymptotic expansion* of $f(x)$, and we write

$$f(x) \cong \sum_{n=0}^{\infty} a_n g_n(x), \tag{10}$$

if

$$f(x) - \sum_{n=0}^{k} a_n g_n(x) = o(g_k(x)) \tag{11}$$

as $x \to x_0$ for every k. Partial sums of the formal series (9) are said to be *asymptotic approximations* to $f(x)$. From relation (11), it is clear that the error associated with an asymptotic approximation to $f(x)$ is little o of the last term retained in the series.

Since the asymptotic sequence $\{g_n(x)\}$ is ordered with respect to the independent variable, the asymptotic expansion (10) is often called a *coordinate expansion*. A convergent Taylor series expansion for $f(x)$ about the point x_0 is the most obvious example of a coordinate expansion. As mentioned in the introduction to this chapter, the coefficients in a Taylor series expansion are uniquely determined. Also, the error associated with using a partial sum of the Taylor series is of the order of the first term neglected. Some simple manipulations will explicitly bring out these two properties for a general coordinate expansion.

Relation (11) may be rewritten in the form

$$f(x) - \sum_{n=0}^{k-1} a_n g_n(x) = a_k g_k(x) + r(x), \tag{12}$$

where

$$r(x) = o(g_k(x)).$$

Thus

$$a_k g_k(x) + r(x) = O(g_k(x)),$$

and hence

$$f(x) - \sum_{n=0}^{k-1} a_n g_n(x) = O(g_k(x)). \tag{13}$$

Since, except possibly at x_0, $g_k(x)$ does not vanish on U, we may divide relation (12) by $g_k(x)$ to obtain

$$a_k = \left\{ \frac{f(x) - \sum_{n=0}^{k-1} a_n g_n(x)}{g_k(x)} \right\} - \frac{r(x)}{g_k(x)} \quad (x \neq x_0).$$

By the definition of little o, $r(x) = o(g_k(x))$ implies

$$\lim_{x \to x_0} \frac{r(x)}{g_k(x)} = 0.$$

Thus

$$a_k = \lim_{x \to x_0} \left\{ \frac{f(x) - \sum_{n=0}^{k-1} a_n g_n(x)}{g_k(x)} \right\}. \tag{14}$$

The limit as $x \to x_0$ above, is necessarily unique. Thus, the coefficients in the formal expansion (9) are uniquely determined and, with respect to a given asymptotic sequence, a given function has a unique asymptotic expansion.

The most commonly used asymptotic sequences are

$$\{h(x)(x - x_0)^n\} \quad \text{as} \quad x \to x_0 \quad (x_0 \text{ finite})$$

and

$$\left\{\frac{h(x)}{x^n}\right\} \quad \text{as} \quad x \to \infty.$$

Asymptotic expansions with respect to these asymptotic sequences are called *asymptotic power series*. Note that

$$f(x) \cong \sum_{n=0}^{\infty} a_n h(x) g_n(x)$$

is equivalent to

$$\frac{f(x)}{h(x)} \cong \sum_{n=0}^{\infty} a_n g_n(x).$$

At this point, the reader should be cautioned that a given function will usually have different asymptotic expansions with respect to different asymptotic sequences. For example, as $x \to \infty$,

$$f(x) = \frac{1}{x + 1} \cong \sum_{n=1}^{\infty} \frac{(-1)^n}{x^n} \tag{15}$$

with respect to the sequence $\{x^{-n}\}$, but with respect to the sequence $\{(x - 1)/x^{2n}\}$, we have

$$f(x) = \frac{1}{x + 1} \cong \sum_{n=1}^{\infty} \frac{x - 1}{x^{2n}}.$$

Also, a given formal series may asymptotically represent more than one function. Since $e^{-x} = o(x^{-n})$ for all n as $x \to \infty$, the asymptotic expansion for $f(x) + e^{-x}$ is

$$g(x) = \frac{1}{x + 1} + e^{-x} \cong \sum_{n=1}^{\infty} \frac{(-1)^n}{x^n}. \tag{16}$$

The functions $f(x)$ in (15) and $g(x)$ in (16) are said to be *asymptotically equivalent*. Stated more generally, two functions $f(x)$ and $g(x)$ are asymptotically equivalent as $x \to x_0$ if

$$\frac{f(x)}{g(x)} \cong 1,$$

that is,

$$f(x) = g(x)\{1 + o(1)\}, \quad \text{as} \quad x \to x_0.$$

1.4 Coordinate Expansions for Functions with Parameter Dependence

Up to this point, we have considered functions which depend on only an independent variable x. We now wish to look at the more general case where the function also depends on a real parameter ϵ.

Let $f(x, \epsilon)$ be a real valued function of a real variable x and a real parameter ϵ. Unless otherwise stated, we will assume ϵ is positive and strictly greater than zero. In most physical applications, ϵ is also usually assumed to be either very large or very small. However, in this section, we shall only assume that values of ϵ lie in the range

$$E : 0 \leq \epsilon_0 < \epsilon < \epsilon_1 \leq +\infty.$$

The results and definitions of Sections 1.2 and 1.3 may be immediately generalized to this case by replacing functions of x by functions of both x and ϵ. We must, however, allow for the possibility that the constants implicit in the order operators might depend on ϵ. For example, the formal series

$$\sum_{n=0}^{\infty} a_n g_n(x, \epsilon)$$

is an asymptotic coordinate expansion of $f(x, \epsilon)$ with respect to the asymptotic sequence $\{g_n(x, \epsilon)\}$ as $x \to x_0$, and we write

$$f(x, \epsilon) \cong \sum_{n=0}^{\infty} a_n g_n(x, \epsilon),$$

if for all k

$$f(x, \epsilon) - \sum_{n=0}^{k} a_n g_n(x, \epsilon) = o(g_k(x, \epsilon))$$

as $x \to x_0$, where the constant implicit in little o might depend on ϵ. If this implicit constant does not depend on ϵ for any k, the coordinate expansion is said to be *uniform* with respect to ϵ.

The most commonly used asymptotic sequences $\{g_n(x, \epsilon)\}$ are "separable" in the sense that each element of the sequence is a function of x times a function of ϵ, i.e.

$$g_n(x, \epsilon) = \phi_n(\epsilon) h_n(x).$$

A natural question to ask is thus:

> Given an asymptotic sequence $\{h_n(x)\}$ as $x \to x_0$ and a sequence of functions $\{\phi_n(\epsilon)\}$ defined for ϵ in E, when is $\{\phi_n(\epsilon) h_n(x)\}$ an asymptotic sequence as $x \to x_0$?

For $\{\phi_n(\epsilon)h_n(x)\}$ to be an asymptotic sequence as $x \to x_0$, by definition, there must be a neighborhood U of x_0 on which none of the elements of the sequence vanish for $x \neq x_0$ and for all n we must have

$$\phi_{n+1}(\epsilon)h_{n+1}(x) = o[\phi_n(\epsilon)h_n(x)] \quad \text{as} \quad x \to x_0,$$

where the constant in the order operator may depend on ϵ.

The nonvanishing of each element of the sequence on U (except possibly at x_0) immediately implies that none of the elements in the sequence $\{\phi_n(\epsilon)\}$ can vanish on E. The order condition implies that we must have

$$\frac{\phi_{n+1}(\epsilon)}{\phi_n(\epsilon)} \cdot \frac{h_{n+1}(x)}{h_n(x)} \to 0 \quad \text{as} \quad x \to x_0. \tag{17}$$

Since $\{h_n(x)\}$ is an asymptotic sequence as $x \to x_0$,

$$\frac{h_{n+1}(x)}{h_n(x)} \to 0 \quad \text{as} \quad x \to x_0.$$

Thus, condition (17) will be satisfied if for all n, $\phi_{n+1}(\epsilon)/\phi_n(\epsilon)$ is bounded for all ϵ in E. If, for all n, this bound is uniform (does not depend on ϵ), then the implicit constant in the order operator will not depend on ϵ. To summarize, we have

> Let $\{\phi_n(\epsilon)\}$ be a sequence of nonvanishing functions defined for ϵ in E such that for all n,
>
> $$\frac{\phi_{n+1}(\epsilon)}{\phi_n(\epsilon)}$$
>
> is bounded on E. If $\{h_n(x)\}$ is an asymptotic sequence as $x \to x_0$, then $\{\phi_n(\epsilon)h_n(x)\}$ is also an asymptotic sequence as $x \to x_0$.

Note that these conditions are trivially satisfied if $\phi(\epsilon)$ is a nonvanishing function defined on E and we let

$$\phi_n(\epsilon) = \phi(\epsilon)$$

for all n.

Suppose $f(x, \epsilon)$ is separable and can be written as

$$f(x, \epsilon) = \phi(\epsilon)h(x),$$

and, with respect to the sequence $\{h_n(x)\}$ as $x \to x_0$, $h(x)$ has an expansion of the form

$$h(x) \cong \sum_{n=0}^{\infty} a_n h_n(x).$$

Then, if $\phi(\epsilon)$ does not vanish, using the above result, $f(x, \epsilon)$ will have an expansion of the form

$$f(x, \epsilon) \cong \phi(\epsilon) \sum_{n=0}^{\infty} a_n h_n(x)$$

with respect to the sequence $\{\phi(\epsilon)h_n(x)\}$ as $x \to x_0$. Since $\phi_n(\epsilon) = \phi(\epsilon)$, so $\phi_{n+1}(\epsilon)/\phi_n(\epsilon) = 1$ for all n, this coordinate expansion will be uniform with respect to ϵ. For example, let

$$f(x, \epsilon) = \frac{\exp(\epsilon)}{x + 1},$$

with $0 < \epsilon < 1$. With respect to the sequence $\{x^{-n}\}$ as $x \to \infty$, we have

$$\frac{1}{x + 1} \cong \sum_{n=1}^{\infty} \frac{(-1)^n}{x^n}.$$

Also, for $0 < \epsilon < 1$, $\exp(\epsilon)$ does not vanish and

$$|\exp(\epsilon)| < e.$$

Thus, as $x \to \infty$ with $0 < \epsilon < 1$,

$$f(x, \epsilon) \cong \exp(\epsilon) \sum_{n=1}^{\infty} \frac{(-1)^n}{x^n}$$

uniformly with respect to ϵ.

1.5 Asymptotic Expansions with Respect to Parameters

In Section 1.4, we considered the behavior of a function $f(x, \epsilon)$ as x went to x_0 for ϵ in E where x_0 was a limit point of I. However, if the dependence of the function on ϵ is analytic, we can also ask for the behavior of $f(x, \epsilon)$ as ϵ goes to ϵ_0 for x in I where ϵ_0 is a limit point of E.

As mentioned previously, in most physical applications the relevant parameter is either very large or very small. To avoid confusion, small parameters will be denoted by ϵ, large parameters by λ. Small and large parameters are, in some sense, interchangeable. For example, if $\epsilon = 1/\lambda^2$, the behavior of a solution $y(x, \epsilon)$ of the differential equation

$$\epsilon y'' - p(x)y = 0$$

for $x \in I$ as $\epsilon \to 0$ in the interval

$$E : 0 < \epsilon < \epsilon_1 << 1$$

is the same as the behavior of the corresponding solution $y(x, \lambda)$ of the differential equation

$$y'' - \lambda^2 p(x)y = 0$$

for $x \in I$ as $\lambda \to +\infty$ in the interval

$$E: 1 << \lambda_0 < \lambda < +\infty.$$

The choice of which of these two problems we wish to work on is usually determined by historical convention. For the methods described in Chapter 2, large parameters are customarily used. On the other hand, in the perturbation methods described in Chapter 3, small parameters are usually used.

We will now deal with expansions with respect to small parameters. Modification needed in the case of large parameters will be discussed briefly at the end of this section.

Let $f(x, \epsilon)$ and $g(x, \epsilon)$ be functions defined for x and ϵ in intervals I and E, respectively. In what follows, let U be an interval of the form

$$U: 0 < \epsilon < \delta < \epsilon_1.$$

Then

(iv) $f(x, \epsilon) = O(g(x, \epsilon))$ if there exists a positive constant C, which may depend on x, such that $|f(x, \epsilon)| \leq C|g(x, \epsilon)|$ for all ϵ in E and x in I.

(v) $f(x, \epsilon) = O(g(x, \epsilon))$ as $\epsilon \to 0$ if there exists a positive constant C, which may depend on x, and an interval U such that $|f(x, \epsilon)| \leq C|g(x, \epsilon)|$ for all ϵ in U and x in I.

(vi) $f(x, \epsilon) = o(g(x, \epsilon))$ as $\epsilon \to 0$ if given any $\mu > 0$, which may depend on x, there exists an interval U such that $|f(x, \epsilon)| < \mu|g(x, \epsilon)|$ for all ϵ in U and x in I.

Note that because of the form of U, which is a consequence of our assumption that $\epsilon > 0$, we have $U \cap E = U$. Thus, the restriction of ϵ to $U \cap E$ in (v) and (vi) is unnecessary. If the constants C and μ do not depend on x, the order estimates are said to be uniform with respect to x.

Let $\{g_n(x, \epsilon)\}$ be an ordered sequence of functions. Then, $\{g_n(x, \epsilon)\}$ is said to be an asymptotic sequence as $\epsilon \to 0$ if there is an interval U on which none of the $g_n(x, \epsilon)$ vanish and for all n

$$g_{n+1}(x, \epsilon) = o(g_n(x, \epsilon)) \quad \text{as} \quad \epsilon \to 0.$$

The most commonly used asymptotic sequence as $\epsilon \to 0$ is $\{\epsilon^n\}$. Expansions with respect to this sequence are also called asymptotic power series. Some other common sequences as $\epsilon \to 0$ are

$$g_n(\epsilon) = \exp\left\{-\frac{1}{\epsilon}\right\}\epsilon^n$$

and

$$g_1(\epsilon) = \log \epsilon, \qquad g_2(\epsilon) = \epsilon \log \epsilon, \qquad g_3(\epsilon) = \epsilon,$$
$$g_4(\epsilon) = \epsilon^2 \log^2 \epsilon, \qquad g_5(\epsilon) = \epsilon^2 \log \epsilon, \qquad g_6(\epsilon) = \epsilon^2, \ldots.$$

As in Section 1.4, these sequences may be combined with suitable sequences $\{\phi_n(x)\}$ to obtain more general asymptotic sequences as $\epsilon \to 0$.

Given a function $f(x, \epsilon)$ and an asymptotic sequence $\{g_n(x, \epsilon)\}$ as $\epsilon \to 0$, the formal series

$$\sum_{n=0}^{\infty} a_n g_n(x, \epsilon) \quad (a_n = \text{constant}),$$

not necessarily convergent, is said to be an asymptotic expansion of $f(x, \epsilon)$ with respect to the parameter ϵ, and we write

$$f(x, \epsilon) \cong \sum_{n=0}^{\infty} a_n g_n(x, \epsilon), \tag{18}$$

if

$$f(x, \epsilon) - \sum_{n=0}^{k} a_n g_n(x, \epsilon) = o(g_k(x, \epsilon))$$

as $\epsilon \to 0$ for every k. Since the asymptotic sequence is now ordered with respect to the parameter, (18) is often called a *parameter expansion*. As with coordinate expansions, a given function has a unique parameter expansion with respect to a given asymptotic sequence. For example, with respect to the asymptotic sequence $\{\epsilon^n\}$ as $\epsilon \to 0$, a unique parameter expansion for the function

$$f(\epsilon) = \frac{1}{1 + \epsilon}$$

is given by the formal series

$$\sum_{n=0}^{\infty} (-1)^n \epsilon^n.$$

However, as before, this formal series is also a unique parameter expansion for the function

$$g(\epsilon) = \frac{1 + \exp\{-1/\epsilon\}}{1 + \epsilon}.$$

Suppose that as $\epsilon \to 0$ a given function $f(x, \epsilon)$ has an expansion of the form

$$f(x, \epsilon) \cong h(x, \epsilon) \sum_{n=0}^{\infty} a_n g_n(x, \epsilon). \tag{19}$$

Further suppose that no $g_n(x, \epsilon)$ contains an exponential factor which grows or decays as $\epsilon \to 0$. Then, in regions where $h(x, \epsilon)$ is exponentially large as $\epsilon \to 0$, we will say that $f(x, \epsilon)$ has *dominant* behavior. In regions where $h(x, \epsilon)$ is exponentially small as $\epsilon \to 0$, we will say that $f(x, \epsilon)$ has *subdominant* behavior. In regions where $h(x, \epsilon)$ is neither exponentially large nor exponentially small as $\epsilon \to 0$, we will say that $f(x, \epsilon)$ is *neutral* or *balanced*. For example if,

$$h(x, \epsilon) = \exp\left\{\frac{x}{\epsilon}\right\},$$

then $f(x, \epsilon)$ is dominant for $x > 0$, subdominant for $x < 0$, and neutral at $x = 0$.

If we wish to work with a large parameter λ rather than a small parameter ϵ, the definitions of order operators, asymptotic sequences, and asymptotic expansions given earlier in this section require only slight modification. Let U be an interval of the form

$$U: \lambda_0 < \delta < \lambda < +\infty.$$

Then, to restate our earlier definition of order operators in terms of λ, we need only use the above interval U, replace ϵ by λ, and replace $\epsilon \to 0$ by $\lambda \to +\infty$. Similarly, a sequence of functions $\{g_n(x, \lambda)\}$ is an asymptotic sequence as $\lambda \to +\infty$ if there is an interval U of the above form on which none of the $g_n(x, \lambda)$ vanish, and for all n

$$g_{n+1}(x, \lambda) = o\{g_n(x, \lambda)\}$$

as $\lambda \to +\infty$. Given a function $f(x, \lambda)$ and an asymptotic sequence $\{g_n(x, \lambda)\}$ as $\lambda \to +\infty$, the formal series

$$\sum_{n=0}^{\infty} a_n g_n(x, \lambda) \qquad (a_n = \text{constant}),$$

not necessarily convergent, is said to be an asymptotic parameter expansion of $f(x, \lambda)$ with respect to λ, and we write

$$f(x, \lambda) \cong \sum_{n=0}^{\infty} a_n g_n(x, \lambda),$$

if

$$f(x, \lambda) - \sum_{n=0}^{k} a_n g_n(x, \lambda) = o\{g_k(x, \lambda)\}$$

as $\lambda \to +\infty$ for all k.

The most commonly used asymptotic sequence as $\lambda \to \infty$ is $\{\lambda^{-n}\}$. Expansions with respect to this sequence are called asymptotic power series.

1.6 Operations with Asymptotic Power Series

Asymptotic power series are the most commonly used asymptotic expansions. For simplicity, we will first restrict ourselves to functions which depend only on x. The limit value x_0 will be taken to be zero and we will consider coordinate expansions with respect to the asymptotic sequence $\{x^n\}$ as $x \to 0$.

Let $f(x)$ and $g(x)$ be two functions with coordinate expansions

$$f(x) \cong \sum_{n=0}^{\infty} a_n x^n \qquad \text{and} \qquad g(x) \cong \sum_{n=0}^{\infty} b_n x^n \tag{20}$$

as $x \to 0$. Then, directly from the definitions, we have

$$f(x) + g(x) \cong \sum_{n=0}^{\infty} (a_n + b_n) x^n \tag{21}$$

and, if c is any constant,

$$cf(x) \cong \sum_{n=0}^{\infty} c a_n x^n \tag{22}$$

as $x \to 0$. It is also easy to show that the product function

$$f \cdot g(x) = f(x) \cdot g(x)$$

has an expansion of the form

$$f(x) \cdot g(x) \cong \sum_{n=0}^{\infty} c_n x^n \tag{23}$$

as $x \to 0$, where

$$c_n = \sum_{j=0}^{n} a_j b_{n-j}.$$

Further, if a_0 is not zero, then

$$\frac{1}{f(x)} \cong \left\{ \sum_{n=0}^{\infty} a_n x^n \right\}^{-1} \tag{24}$$

Asymptotic power series of functions of both x and ϵ with respect to the asymptotic sequence $\{\phi_n(\epsilon) x^n\}$ as $x \to 0$ obviously satisfy relations like (21) and (22) with $f(x)$ and $g(x)$ replaced by $f(x, \epsilon)$ and $g(x, \epsilon)$ and x^n replaced by $\phi_n(\epsilon) x^n$. If $\phi_n(\epsilon) = \phi(\epsilon)$ for all n, then the multiplication and division properties also hold. However, if $\phi_n(\epsilon) \neq \phi(\epsilon)$ for all n, further restrictions may have to be imposed on elements of the sequence $\{\phi_n(\epsilon)\}$. These same results are also true for parameter expansions of functions of x and ϵ with respect to the asymptotic sequence $\{\phi_n(x)\epsilon^n\}$ as $\epsilon \to 0$.

Again, consider functions which depend only on x and have coordinate expansions as in (20). The reader should be cautioned that in working with asymptotic approximations rather than full asymptotic series, great care must be used. For example, using the asymptotic approximations

$$f(x) = \sum_{n=0}^{p} a_n x^n + O(x^{p+1})$$

and

$$g(x) = \sum_{n=0}^{q} b_n x^n + O(x^{q+1})$$

as $x \to 0$, if $p < q$ we have

$$f(x) + g(x) = \sum_{n=0}^{p} (a_n + b_n)x^n + O(x^{p+1})$$

and

$$f(x) \cdot g(x) = \sum_{n=0}^{p} c_n x^n + O(x^{p+1})$$

as $x \to 0$. The terms

$$B_{p,q}(x) = b_{p+1}x^{p+1} + \cdots + b_q x^q$$

must be ignored in the approximation to $g(x)$, since

$$B_{p,q}(x) + O(x^{p+1}) = O(x^{p+1})$$

and

$$a_0 \cdot B_{p,q}(x) = O(x^{p+1})$$

as $x \to 0$.

We now wish to consider the asymptotic power series for the integral and derivative of $f(x)$ as $x \to 0$. Suppose that for sufficiently small values of x,

$$F(x) = \int_0^x f(t) \, dt$$

exists. Then, if $f(x)$ has an expansion as in (20), $F(x)$ also has a power series expansion as $x \to 0$ and this expansion may be obtained by integrating the expansion for $f(x)$ term by term. Thus

$$F(x) \cong \sum_{n=0}^{\infty} \frac{a_n}{n+1} x^{n+1} \quad \text{as} \quad x \to 0.$$

Differentiation, on the other hand, is a much more difficult problem. For one thing, $f'(x)$ may not even exist. Even if it does exist, $f'(x)$ may not have an asymptotic power series expansion. However, if it is known that $f'(x)$ exists and also possesses an asymptotic power series expansion, then it can be shown that this expansion can be uniquely obtained from the expansion for $f(x)$ by differentiating term by term, i.e.

$$f'(x) \cong \sum_{n=1}^{\infty} n a_n x^{n-1} \qquad \text{as } x \to 0.$$

If x_0 is finite but nonzero, the above results hold if we transform to a new independent variable

$$\tilde{x} = x - x_0.$$

However, if $x_0 = +\infty$ and we deal with power series expansions of the form

$$f(x) \cong \sum_{n=0}^{\infty} \frac{a_n}{x^n}$$

as $x \to \infty$, additional care must be taken. This case is dealt with in the problems at the end of this section.

Problems

1. Show that $(x+1)^x = e x^x \left\{ 1 + O\left(\frac{1}{x}\right) \right\}$ as $x \to \infty$.

2. Obtain two terms in the asymptotic expansion of the integral

    $$\int_x^{\infty} e^{-t^2} \, dt$$

    as $x \to \infty$. What is the order of magnitude of the error for this second approximation? Show that the error in stopping at the first term is less in absolute value than the first term neglected. (*Hint*: Integrate by parts.)

3. For x in a bounded interval I which does not include $x = 0$, let $f(x, \epsilon)$ have an asymptotic parameter expansion

    $$f(x, \epsilon) \cong \exp\left\{\frac{x}{\epsilon}\right\} P(x, \epsilon)$$

    as $\epsilon \to 0$ where $P(x, \epsilon)$ is a power series in ϵ having coefficients which depend only on x, e.g.,

    $$P(x, \epsilon) = \sum_{n=0}^{\infty} g_n(x) \epsilon^n.$$

Suppose each element of the sequence $\{g_n(x)\}$ is a slowly varying, differentiable function on I. Then, the parameter expansion for the derivative $f'(x, \epsilon)$ is given by

$$f'(x, \epsilon) \cong \frac{1}{\epsilon} \exp\left\{\frac{x}{\epsilon}\right\} [P(x, \epsilon) + \epsilon P'(x, \epsilon)]$$

as $\epsilon \to 0$. Assume that we know only the first approximation,

$$f(x, \epsilon) = g_0(x) \exp\left\{\frac{x}{\epsilon}\right\} \{1 + O(\epsilon)\}.$$

Show that the first approximation to $f'(x, \epsilon)$ is given by

$$f'(x, \epsilon) = \frac{1}{\epsilon} g_0(x) \exp\left\{\frac{x}{\epsilon}\right\} \{1 + O(\epsilon)\}$$

and may be obtained by differentiating only the rapidly varying exponential in the first approximation to $f(x, \epsilon)$, treating the relatively slowly varying portion $g_0(x)$ as a constant.

4. Suppose that, as $x \to \infty$,

$$f(x) \cong \sum_{n=0}^{\infty} \frac{a_n}{x^n} \qquad \text{and} \qquad g(x) \cong \sum_{n=0}^{\infty} \frac{b_n}{x^n}.$$

Show that, as $x \to \infty$,

$$f(x) + g(x) \cong \sum_{n=0}^{\infty} \frac{(a_n + b_n)}{x^n}$$

$$f(x) \cdot g(x) \cong \sum_{j=0}^{\infty} \frac{c_n}{x^n}$$

where

$$c_n = \sum_{j=0}^{n} a_j b_{n-j}.$$

If $a_0 \neq 0$, show

$$\frac{1}{f(x)} \cong \frac{1}{a_0} + \sum_{n=1}^{\infty} \frac{d_n}{x^n}.$$

5. Suppose that for x suitably large, $f(x)$ is continuous and, as $x \to \infty$,

$$f(x) \cong \sum_{n=0}^{\infty} \frac{a_n}{x^n}.$$

Show that as $x \to \infty$

$$\int_x^{\infty} \left\{ f(t) - a_0 - \frac{a_1}{t} \right\} dt \cong \sum_{n=1}^{\infty} \frac{a_{n+1}}{n x^n}.$$

Use this result to show that if, for sufficiently large $x, f(x)$ has a continuous derivative $f'(x)$ and as $x \to \infty$

$$f'(x) \cong \sum_{n=0}^{\infty} \frac{b_n}{x^n}.$$

Then

$$b_0 = 0, \qquad b_1 = 0,$$

and for $n \geq 2$

$$b_n = -(n-1)a_{n-1}.$$

References

1. E. T. Copson, *Asymptotic Expansions*, University Press, Cambridge, 1965.
2. H. G. De Bruijn, *Asymptotic Methods in Analysis*, North-Holland Publishing Company, Amsterdam, 1958.
3. A. Erdelyi, *Asymptotic Expansions*, Dover, New York, 1956.
4. H. Jeffreys, *Asymptotic Approximations*, Clarendon Press, Oxford, 1962.
5. R. Courant and D. Hilbert, *Methods of Mathematical Physics*, Volume 1, Interscience, New York, 1953.

2

Asymptotic Approximations
for Second Order Equations

2.1 Introduction

The asymptotic behavior of solutions of second order ordinary differential equations containing a real parameter λ will be considered in this chapter. Following convention, λ will be assumed to be large and positive. In Section 2.2, it is shown that by means of a simple transformation a general second order equation can be transformed to a form having no first derivative term. For this reason, only equations of the form

$$w'' - p(x, \lambda)w = 0$$

will be considered.

Let $p(x, \lambda)$ be a real valued function defined for x in an interval I and for λ in an interval E. If x_0 is a point in I and, for fixed λ, $p(x, \lambda)$ is analytic in a neighborhood of x_0, then formal solutions of the differential equation may be obtained in the form of convergent power series in $(x - x_0)$ with coefficients depending on λ. However, very often the behavior of solutions in a neighborhood of x_0 for fixed λ is not of primary interest. This is the case in many physical applications. Instead, for x in I, we wish to know the asymptotic behavior of the solutions as λ approaches some limiting value λ_0. This is the problem considered in this chapter. In particular, the limiting value λ_0 will be taken to be $+\infty$.

As might be expected, the asymptotic behavior of the solutions and the methodology used to obtain approximations depends on the behavior of $p(x, \lambda)$. In particular, if r is a rational number and, as $\lambda \to \infty$, $p(x, \lambda)$ has an asymptotic parameter expansion of the form

$$p(x, \lambda) \cong \lambda^r \sum_{n=0}^{\infty} p_n(x) \lambda^{-rn/2},$$

then subintervals I' of I containing a point where $p_0(x)$ vanishes must be treated differently from subintervals I' on which $p_0(x)$ does not vanish. Singularities of $p_0(x)$ must also be taken into account.

The behavior of solutions on subintervals containing a simple zero of $p_0(x)$ is treated at the end of this chapter. Earlier sections deal with subintervals on which $p_0(x)$ is either strictly positive or strictly negative. This case is treated through both the straightforward WKB method and the comparison equation–construction technique. Strict error bounds for first approximations are also derived.

2.2 Normal Form

Let $\phi_1(x)$ and $\phi_2(x)$ be two linearly independent solutions on an interval I of the second order ordinary differential equation

$$y'' + q(x)y' + r(x)y = 0. \tag{1}$$

The *Wronskian* $W(x)$ of $\phi_1(x)$ and $\phi_2(x)$ is defined by the determinant

$$W(x) = \begin{vmatrix} \phi_1(x) & \phi_2(x) \\ \phi_1'(x) & \phi_2'(x) \end{vmatrix},$$

i.e.,

$$W(x) = \phi_1(x)\phi_2'(x) - \phi_1'(x)\phi_2(x).$$

Since $\phi_1(x)$ and $\phi_2(x)$ are linearly independent solutions of Equation (1), $W(x)$ is nonzero for all x in I. Also, if x_0 is any point in I, $W(x)$ may be obtained from $W(x_0)$ through the relation

$$W(x) = W(x_0) \exp\left\{ -\int_{x_0}^{x} q(t)\, dt \right\}. \tag{2}$$

Thus, if $q(x) \equiv 0$ in (1), then the Wronskian $W(x)$ is constant on I and

$$W(x) = W(x_0) = \phi_1(x_0)\phi_2'(x_0) - \phi_1'(x_0)\phi_2(x_0).$$

Consider now the inhomogeneous differential equation

$$y'' + q(x)y' + r(x)y = f(x). \tag{3}$$

If $\phi_1(x)$ and $\phi_2(x)$ are again linearly independent solutions of the homogeneous equation (1), then the general solution of (3) must be of the form

$$\phi(x) = \phi_p(x) + c_1\phi_1(x) + c_2\phi_2(x),$$

where c_1 and c_2 are arbitrary constants and $\phi_p(x)$ is a particular solution of the form

$$\phi_p(x) = \int_{x_0}^{x} \frac{[\phi_1(t)\phi_2(x) - \phi_1(x)\phi_2(t)]f(t)}{W(t)}\,dt. \tag{4}$$

Because $W(t)$ appears in the denominator, if $W(x)$ is not constant on I the integrand in (4) may be quite complicated. However, if $q(x) \equiv 0$, then $W(t)$ is a constant and may be taken outside the integral. Since such simplifications are highly desirable, it is natural to ask if Equation (3) can be transformed to a form

$$w'' - p(x)w = g(x), \tag{5}$$

in which there is no first derivative term. Equation (5) is said to be in *normal form*.

To make this transformation, let

$$y(x) = w(x)\exp\left\{-\frac{1}{2}\int^{x} q(t)\,dt\right\}. \tag{6}$$

Then, Equation (3) transforms into Equation (5) with

$$p(x) = -r(x) + \tfrac{1}{4}q^2(x) + \tfrac{1}{2}q'(x)$$

and

$$g(x) = f(x)\exp\left\{\frac{1}{2}\int^{x} q(t)\,dt\right\}.$$

Since this transformation can always be made if $q(x)$ is continuous, in the remainder of this chapter we will deal only with differential equations in normal form.

2.3   The WKB Approximation

Let $p(x, \lambda)$ be a real valued function of a real variable x and a real parameter λ defined for x in the open interval

$$I : a < x < b.$$

The constants a and b may be $+\infty$ and $-\infty$, respectively. Assume that as $\lambda \to +\infty$, the function $p(x, \lambda)$ has an asymptotic parameter expansion of the form

$$p(x, \lambda) \cong \lambda^r \sum_{n=0}^{\infty} p_n(x) \lambda^{-rn/2}, \tag{7}$$

where r is a rational number and the functions $p_n(x)$ are defined and sufficiently differentiable on I. In the remainder of this chapter, we will seek the asymptotic behavior of solutions $w(x)$ of the second order differential equation

$$w'' - p(x, \lambda)w = 0 \tag{8}$$

as $\lambda \to \infty$ for x in I.

In this section, we will consider the special case

$$p(x, \lambda) = \lambda^2 p(x)$$

and work on subintervals I' of I on which $p(x)$ does not vanish. The function $p(x)$ may have double poles at points $a_1, \ldots, a_m$ in I', e.g.,

$$p(x) = \frac{q(x)}{(x - a_1)^2 \cdots (x - a_m)^2},$$

where $q(x)$ does not vanish on I', but no other types of singularities are allowable. The method that will be used in this case to obtain the asymptotic behavior of $w(x)$ as $\lambda \to \infty$ is usually called the WKB approximation (after Wentzel, Kramers, and Brillouin).

For the moment, $p(x)$ will be assumed strictly greater than zero on the subinterval I' under consideration. The case where $p(x)$ is strictly less than zero will be treated at the end of this section. The first step in the WKB method is to seek a solution in exponential form. Let

$$w(x) = \exp\left\{ \int^x \phi(t)dt \right\} = E(x), \tag{9}$$

i.e.,

$$\phi(x) = \frac{w'(x)}{w(x)}.$$

Then

$$w'(x) = \phi(x)E(x)$$

and

$$w''(x) = \{\phi^2(x) + \phi'(x)\}E(x).$$

Substituting these results into (8), we obtain a first order nonlinear *Ricatti equation*

$$\phi' + \phi^2 = \lambda^2 p. \tag{10}$$

Since (10) suggests that $\phi(x)$ must be large like λ, assume that $\phi(x)$ has an expansion of the form

$$\phi(x) \cong \lambda \sum_{n=0}^{\infty} \phi_n(x) \lambda^{-n}. \qquad (11)$$

Thus

$$\phi'(x) \cong \lambda \sum_{n=0}^{\infty} \phi_n'(x) \lambda^{-n}$$

$$\phi^2(x) \cong \lambda^2 \sum_{n=0}^{\infty} \psi_n(x) \lambda^{-n},$$

where

$$\psi_n(x) = \sum_{j=0}^{n} \phi_j(x) \phi_{n-j}(x).$$

Using these expansions in (10), we find that

$$\lambda^2 (\psi_0 - p) + \sum_{n=1}^{\infty} \{\phi_{n-1}' + \psi_n\} \lambda^{2-n} = 0.$$

Equating powers of λ to zero, and using the fact that $\psi_0(x) = \phi_0^2(x)$ and $\psi_1(x) = 2\phi_0(x)\phi_1(x)$, we obtain

$$\phi_0^2 - p = 0,$$

and

$$\phi_0' + 2\phi_0 \phi_1 = 0$$

while, for $n \geq 2$,

$$\phi_{n-1}' + 2\phi_0 \phi_n + \sum_{j=1}^{n-1} \phi_j \phi_{n-j} = 0.$$

These relations may be solved successively for the coefficients in the expansion of $\phi(x)$. As is obvious from the above, the method does not involve solving differential equations. At each stage, $\phi_n(x)$ is obtained algebraically.

In most applications, only the first two terms are retained in the expansion (11) for $\phi(x)$. Solving explicitly for $\phi_0(x)$ and $\phi_1(x)$, we find

$$\phi_0(x) = \pm \{p(x)\}^{1/2} \qquad (12)$$

and

$$\phi_1(x) = -\frac{1}{2} \frac{\phi_0'(x)}{\phi_0(x)}.$$

Note that

$$\int^x \phi_1(t)\,dt = -\frac{1}{2} \int^x \frac{\phi_0'(t)}{\phi_0(t)}\,dt$$

$$= -\frac{1}{2} \ln |\phi_0(x)|.$$

Thus

$$\exp\left\{\int^x \phi_1(t)\,dt\right\} = |\phi_0(x)|^{-1/2} \tag{13}$$

$$= |p(x)|^{-1/4}.$$

Having obtained the expansion for $\phi(x)$, we can now use the transformation (9) to obtain the behavior of $w(x)$ as $\lambda \to \infty$. We find

$$w(x) = \exp\left\{\int^x \left[\lambda\phi_0(t) + \phi_1(t) + O\left(\frac{1}{\lambda}\right)\right] dt\right\}$$

or

$$w(x) = \exp\left\{\pm\lambda \int^x \{p(t)\}^{1/2}\,dt\right\}$$

$$\cdot \exp\left\{\int^x \phi_1(t)\,dt\right\}\left\{1 + O\left(\frac{1}{\lambda}\right)\right\}.$$

Since $p(x)$ is positive on I', $|p(x)| = p(x)$ and we have

$$w(x) = \{p(x)\}^{-1/4} \exp\left\{\pm\lambda \int^x \{p(t)\}^{1/2}\,dt\right\}$$

$$\cdot \left\{1 + O\left(\frac{1}{\lambda}\right)\right\}. \tag{14}$$

Note that the rapidly varying exponential portion comes from the $\phi_0(x)$ term while the factor $\{p(x)\}^{-1/4}$, which is slowly varying compared to the exponential, comes from the $\phi_1(x)$ term. The restriction that $p(x) > 0$ on I' is required to make the one-half and one-fourth roots of $p(x)$ in (14) real and $\{p(x)\}^{-1/4}$ well behaved.

Let $w_1(x)$ and $w_2(x)$ denote the two solutions of the differential equation

$$w'' - \lambda^2 p(x)w = 0,$$

whose first approximations on intervals I' where $p(x) > 0$ are given by relation (14) using the plus and minus signs in the exponential, respectively. These two solutions are linearly independent. Therefore, any general solution of the differential equation can be written as a linear combination of $w_1(x)$ and $w_2(x)$, where the constants in the linear combination may depend on λ. Thus, if $w(x)$ is any solution of the above differential equation, then

$$w(x) = c_1(\lambda)w_1(x) + c_2(\lambda)w_2(x).$$

We will now deal briefly with approximations valid on subintervals I' on which $p(x) < 0$. The above method goes through with only minor modifications in this case, although the behavior of the solutions is completely changed. Using the transformation (9) and expansion (11), we find that

$$\phi_0(x) = \pm \{p(x)\}^{1/2} = \pm i\{-p(x)\}^{1/2},$$

where $i = \sqrt{-1}$ and $-p(x) > 0$ on I'. Also, in (13), since we have an absolute value, we must take the exponential of the integral equal to $\{-p(x)\}^{-1/4}$. Thus, on subintervals I' on which $p(x) < 0$

$$w(x) = \{-p(x)\}^{-1/4} \exp\left\{\pm i\lambda \int^x \{-p(t)\}^{1/2} \, dt\right\} \cdot \left\{1 + O\left(\frac{1}{\lambda}\right)\right\}$$

and the approximations involve complex exponentials.

Let $u_1(x)$ and $u_2(x)$ be the two linearly independent, complex valued solutions whose first approximations on I' are given by the above expression using the plus and minus sign in the exponential, respectively. We will use the fact that any linear combination of $u_1(x)$ and $u_2(x)$ is also a solution to obtain a second pair of linearly independent solutions whose approximations on I' involve only real valued functions. In particular, let a be a real constant, c a possibly complex constant, and define

$$w_1(x) = \frac{c}{2i} \{e^{ia}u_1(x) - e^{-ia}u_2(x)\},$$

$$w_2(x) = \frac{c}{2} \{e^{ia}u_1(x) + e^{-ia}u_2(x)\}.$$

Then, $w_1(x)$ and $w_2(x)$ are two linearly independent solutions of the differential equation such that on the subinterval I' on which $p(x) < 0$

$$w_1(x) = c\{-p(x)\}^{-1/4} \sin\left\{\lambda \int^x \{-p(t)\}^{1/2} \, dt + a\right\} \cdot \left\{1 + O\left(\frac{1}{\lambda}\right)\right\}$$

$$\tag{15}$$

and

$$w_2(x) = c\{-p(x)\}^{-1/4} \cos\left\{\lambda \int^x \{-p(t)\}^{1/2} \, dt + a\right\} \cdot \left\{1 + O\left(\frac{1}{\lambda}\right)\right\}.$$

The constant a is called a *phase angle*.

The WKB approximation is one of the most commonly used asymptotic methods since in many physical applications $p(x, \lambda)$ has the special form $\lambda^2 p(x)$. Accordingly, it is worthwhile to summarize the above results.

Let $p(x)$ be defined for x in an interval I, and let λ be a large, positive, real parameter. Then, $w_1(x)$ and $w_2(x)$ are two linearly independent solutions of the differential equation

$$w'' - \lambda^2 p(x)w = 0$$

having the following asymptotic properties:

(1) On subintervals I' of I on which $p(x) > 0$, as $\lambda \to \infty$

$$w_1(x) = \{p(x)\}^{-1/4} \exp\left\{\lambda \int^x \{p(t)\}^{1/2}\, dt\right\}$$
$$\cdot \left\{1 + O\left(\frac{1}{\lambda}\right)\right\}$$

and

$$w_2(x) = \{p(x)\}^{-1/4} \exp\left\{-\lambda \int^x \{p(t)\}^{1/2}\, dt\right\}$$
$$\cdot \left\{1 + O\left(\frac{1}{\lambda}\right)\right\}.$$

(2) On subintervals I' of I on which $p(x) < 0$, as $\lambda \to \infty$

$$w_1(x) = c\{-p(x)\}^{-1/4}$$
$$\cdot \sin\left\{\lambda \int^x \{-p(t)\}^{1/2}\, dt + a\right\}$$
$$\cdot \left\{1 + O\left(\frac{1}{\lambda}\right)\right\}$$

and

$$w_2(x) = c\{-p(x)\}^{-1/4}$$
$$\cdot \cos\left\{\lambda \int^x \{-p(t)\}^{1/2}\, dt + a\right\}$$
$$\cdot \left\{1 + O\left(\frac{1}{\lambda}\right)\right\},$$

*where the phase angle a is a real constant and c
is a possibly complex constant. Thus, if on I',
$p(x) > 0$, the approximations are exponentially
large or exponentially small. If $p(x) < 0$ on
I', the approximations are oscillatory. Again,
any solution of the differential equation can be
written as a linear combination of $w_1(x)$ and
$w_2(x)$ where the constants in the linear combination may depend on λ.*

Example

As will be seen at the end of this chapter, the Airy functions
play a central role in obtaining approximations to the solutions
of Equation (8) valid in intervals where $p_0(x)$ has a simple
zero. For purposes of definition, a complex variable z must be
introduced.

The Airy functions $P_j(z)$ ($j = 1, 2, 3$) are defined in terms of
the contour integrals

$$P_j(z) = \frac{1}{2\pi i} \int_{C_j} \exp\left\{ zt - \frac{t^3}{3} \right\} dt, \qquad (16)$$

where the paths of integration C_j in the complex t-plane are
shown in Figure 2.1.

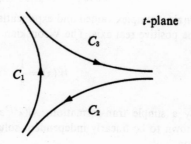

Paths of integration for The Airy Functions $P_j(z)$.

Figure 2.1

$P_1(z)$ is usually denoted by $Ai(z)$. Since the paths of integration
C_2 and C_3 may be obtained from C_1 by rotating C_1 through an

angle $\pm 2\pi/3$, from (16) $P_2(z)$ and $P_3(z)$ are related to $Ai(z)$ in the manner

$$P_2(z) = e^{2\pi i/3} Ai(ze^{2\pi i/3}) \qquad (17)$$

and

$$P_3(z) = e^{-2\pi i/3} Ai(ze^{-2\pi i/3}).$$

Also, since the integrand in (16) has no singularities inside the closed path $C_1 + C_2 + C_3$, Cauchy's theorem gives the further relation

$$Ai(z) + P_2(z) + P_3(z) = 0,$$

which is called an *exact connection formula*. The Airy function $P_j(z)$ has the property that it is real on the ray

$$R_j : z = |z|e^{i\theta_j},$$

with $\theta_1 = 0$, $\theta_2 = 4\pi/3$, and $\theta_3 = 2\pi/3$.

Also, if z is on the ray R_j and $|z|$ is large, then $P_j(z)$ is exponentially small.

By direct differentiation of (16), the Airy functions are found to be exact solutions of the differential equation

$$y'' - zy = 0.$$

Any two of the three Airy functions are a set of linearly independent solutions of this equation. However, the most numerically satisfactory set of linearly independent solutions are the functions $Ai(z)$ and $Bi(z)$ where

$$Bi(z) = i\{P_3(z) - P_2(z)\}.$$

$Bi(z)$ is complex valued and exponentially large for large $|z|$ on the positive real axis. The Wronskian $W(x)$ of $Ai(z)$ and $Bi(z)$ is

$$W(x) = \frac{1}{\pi}.$$

By a simple transformation, $Ai(\lambda^{2/3}z)$ and $Bi(\lambda^{2/3}z)$ can be shown to be linearly independent solutions of the Airy equation

$$w'' - \lambda^2 zw = 0.$$

If λ is a real parameter, then $Ai(\lambda^{2/3}z)$ is real on the positive real axis. Further, since λ is assumed large, $Ai(\lambda^{2/3}z)$ is exponentially small on the real axis if $z = O(1)$. We shall now use these two facts to obtain approximations to $Ai(\lambda^{2/3}x)$ where x is again a real variable.

By the above discussion, $Ai(\lambda^{2/3}x)$ is a solution of the differential equation with real coefficients

$$w'' - \lambda^2 xw = 0. \tag{18}$$

Applying the previous theorem with $p(x) = x$, we must thus have that

$$Ai(\lambda^{2/3}x) = c_1(\lambda)w_1(x) + c_2(\lambda)w_2(x)$$

and, since the approximation to $w_1(x)$ is exponentially large for $0 < x = O(1)$, we must choose $c_1(\lambda) = 0$. Unfortunately, $c_2(\lambda)$ cannot be determined directly from (18). This is because we are asking for much more than simply approximations to a solution of (18) which is exponentially small for $0 < x = O(1)$. Instead, we want approximations to a *specific* function, namely $Ai(\lambda^{2/3}z)$, which is a solution of (18). In such cases, we expect to have to use information not directly related to the differential equation.

For this problem, we not only have to determine $c_2(\lambda)$, but also the phase angle a and the constant c which appear in the approximation to $w_2(x)$ for $0 > x = O(1)$. The value of $c_2(\lambda)$ may be found by comparing the approximation to $w_2(x)$ valid for $0 < x = O(1)$ with the expansion for $Ai(\lambda^{2/3}x)$ obtained by the classical method of steepest descents which works with the integral (16) directly. We find that

$$c_2(\lambda) = \tfrac{1}{2}\pi^{-1/2}\lambda^{-1/6}.$$

To determine the constants a and c, a complex variable z must be reintroduced. We may then use relations (17) and the exact connection formula to find a and c. Since this procedure is really beyond the scope of this book, we will only state the final result

$$a = -\pi/4 \qquad \text{and} \qquad c = 2e^{-\pi i/4}.$$

Thus, for $x = O(1)$ on the positive real axis

$$Ai(\lambda^{2/3}x) = \tfrac{1}{2}\pi^{-1/2}\lambda^{-1/6}x^{-1/4}\exp\{-\tfrac{2}{3}\lambda x^{3/2}\} \cdot \left\{1 + O\left(\frac{1}{\lambda}\right)\right\}, \tag{19}$$

while for $x = O(1)$ on the negative real axis

$$Ai(\lambda^{2/3}x) = \pi^{-1/2}\lambda^{-1/6}x^{-1/4}\cos\left\{\frac{2}{3}\lambda(-x)^{3/2} - \frac{\pi}{4}\right\} \cdot \left\{1 + O\left(\frac{1}{\lambda}\right)\right\}. \tag{20}$$

In most problems where the WKB method is applied, solutions of the differential equation

$$w'' - \lambda^2 p(x)w = 0$$

are unknown and we merely want approximations to a solution of this equation which is exponentially small, say, when $p(x) > 0$. In this case, the constants $c_2(\lambda)$, a, and c can be determined directly. This question will be dealt with in Section 9 of this chapter when we consider the connection problem.

2.4 A More General Case

The arguments of the preceding section will now be generalized to treat the equation

$$w'' - p(x, \lambda)w = 0,$$

where $p(x, \lambda)$ has an expansion as in (7). We will work on subintervals I' of I on which $p_0(x)$ does not vanish. The function $p_0(x)$ will first be assumed strictly positive on I' and may have double poles.

Making the transformation

$$w(x) = \exp\left\{\int^x \phi(t)\,dt\right\}$$

and using the expansion for $p(x, \lambda)$, our original differential equation becomes

$$\phi' + \phi^2 = \lambda^r \sum_{n=0}^{\infty} p_n(x)\lambda^{-rn/2}. \tag{21}$$

We will assume that as $\lambda \to \infty$ for x in I', $\phi(x)$ has an asymptotic parameter expansion of the form

$$\phi(x) \cong \lambda^{r/2} \sum_{n=0}^{\infty} \phi_n(x)\lambda^{-rn/2}. \tag{22}$$

With $\psi_n(x)$ as in Section 2.3, we find that

$$\phi^2(x) \cong \lambda^r \sum_{n=0}^{\infty} \psi_n(x)\lambda^{-rn/2}$$

and

$$\phi'(x) \cong \lambda^{r/2} \sum_{n=0}^{\infty} \phi'_n(x)\lambda^{-rn/2}.$$

Using these expansions, Equation (21) may be written as

$$\lambda^r(\psi_0 - p_0) + \sum_{n=1}^{\infty} (\phi'_{n-1} + \psi_n - p_n)\lambda^{(r/2)(2-n)} = 0.$$

Equating powers of λ and using the expression for $\psi_n(x)$ we thus find

$$\phi_0^2 = p_0,$$

$$\phi'_0 + 2\phi_0\phi_1 = p_1,$$

and, for $n \geq 2$,

$$\phi'_{n-1} + 2\phi_0\phi_n + \sum_{j=1}^{n-1} \phi_j\phi_{n-j} = p_n.$$

Solving the first two relations for $\phi_0(x)$ and $\phi_1(x)$ gives

$$\phi_0(x) = \pm\{p_0(x)\}^{1/2}$$

and

$$\phi_1(x) = \frac{d}{dx}\{\ln[p_0(x)]^{-1/4}\} \pm \tfrac{1}{2}p_1(x)\{p_0(x)\}^{-1/2}.$$

Our original transformation thus gives

$$w(x) \cong \{p_0(x)\}^{-1/4}$$
$$\cdot \exp\left\{\pm \int^x [\lambda(p_0(t))^{1/2} + \tfrac{1}{2}p_1(t)(p_0(t))^{-1/2}]\, dt\right\}$$
$$\cdot \exp\left\{\sum_{n=2}^{\infty} \lambda^{(r/2)(1-n)} \int^x \phi_n(t)\, dt\right\}$$

or, if we let

$$\tilde{w} = \{p_0(x)\}^{-1/4} \cdot \exp\left\{\pm \int^x [\lambda(p_0(t))^{1/2} + \tfrac{1}{2}p_1(t)(p_0(t))^{-1/2}]\, dt\right\},$$

$$(23)$$

we have

$$w(x) = \tilde{w}(x)\left\{1 + O\left(\frac{1}{\lambda}\right)\right\}. \tag{24}$$

If $p_0(x)$ is strictly negative on I', we have the same sort of situation as in Section 2.3. The function $p_0(x)$ is replaced by $|p_0(x)| = -p_0(x)$ and the approximations involve sines and cosines rather than exponentials. Hence, the behavior of the approximations is oscillatory rather than exponentially growing or decaying.

Example

Consider the *Bessel equation*

$$w'' + \left\{\frac{\lambda^2}{4x} - \frac{\lambda^2 - 1}{4x^2}\right\}w = 0. \tag{25}$$

Two linearly independent solutions of this equation involve the Bessel function of the first kind $J_\lambda(\lambda x^{1/2})$ and the Bessel function of the second kind $Y_\lambda(\lambda x^{1/2})$. If we let C_λ stand for either J_λ or Y_λ, then exact solutions of Equation (25) are

$$w(x) = x^{1/2}C_\lambda(\lambda x^{1/2}).$$

This fact, and some knowledge of the behavior of the Bessel functions in a neighborhood of $x = 0$, will be used to find the behavior of $J_\lambda(\lambda x^{1/2})$ as $\lambda \to \infty$ with x in the interval, $I' : 0 < x < 1$.

We wish to apply our method with

$$r = 2$$

and

$$p_0(x) = \frac{1}{4x^2}(1 - x), \qquad p_1(x) = 0,$$

$$p_2(x) = -\frac{1}{4x^2}, \qquad \text{and} \qquad p_j(x) = 0 \quad (j \geq 3).$$

On the interval I', $p_0(x)$ is strictly positive. By direct calculation,

$$\phi_0(x) = \pm\frac{(1 - x)^{1/2}}{2x},$$

$$\phi_1(x) = \frac{d}{dx}\{\ln[(2x)^{1/2}(1 - x)^{-1/4}]\},$$

and

$$\phi_2(x) = \mp\frac{x + 2}{4(1 - x)^{5/2}}.$$

Since

$$\int^x \phi_0(t)\,dt = (1 - x)^{1/2} + \ln\left\{\frac{1 - (1 - x)^{1/2}}{1 + (1 - x)^{1/2}}\right\}^{1/2},$$

relation (23) yields

$$\tilde{w}(x) = (2x)^{1/2}(1 - x)^{-1/4}\left\{\frac{1 - (1 - x)^{1/2}}{1 + (1 - x)^{1/2}}\right\}^{\pm\lambda/2} \cdot \exp\{\pm\lambda(1 - x)^{1/2}\}.$$

$$(26)$$

Consider the solution $w_1(x)$ given by

$$w_1(x) = \tilde{w}_1(x)\left\{1 + O\left(\frac{1}{\lambda}\right)\right\} \qquad \text{as} \qquad \lambda \to \infty,$$

where $\tilde{w}_1(x)$ corresponds to using the plus signs in relation (26). Since $x^{1/2}J(\lambda x^{1/2})$ and $x^{1/2}Y_\lambda(\lambda x^{1/2})$ are linearly independent solutions of Equation (25), we must have

$$x^{-1/2}w_1(x) = c_1(\lambda)J_\lambda(\lambda x^{1/2}) + c_2(\lambda)Y_\lambda(\lambda x^{1/2}). \qquad (27)$$

We will show that $x^{-1/2}w_1(x)$ must be a multiple of $J_\lambda(\lambda x^{1/2})$.

As $x \to 0$ for λ fixed, the Bessel functions have the asymptotic behavior

$$J_\lambda(\lambda x^{1/2}) \cong \frac{(\lambda x^{1/2}/2)^\lambda}{\Gamma(\lambda + 1)} \qquad (28)$$

and

$$Y_\lambda(\lambda x^{1/2}) \cong -\frac{\Gamma(\lambda)}{\pi}\left(\frac{\lambda x^{1/2}}{2}\right)^{-\lambda}, \qquad (29)$$

where $\Gamma(\lambda)$ is the *Gamma function* defined by

$$\Gamma(n) = \int_0^\infty t^{n-1}e^{-t}\,dt$$

and

$$\Gamma(n) = (n-1)!$$

To find $c_1(\lambda)$ and $c_2(\lambda)$, we obtain the behavior of $w_1(x)$ as $x \to 0$ for fixed λ and then use (27), (28), and (29).

As $x \to 0$,

$$1 - (1-x)^{1/2} = \tfrac{1}{2}x + O(x^2),$$
$$1 + (1-x)^{1/2} = 2 - \tfrac{1}{2}x + O(x^2),$$

and hence

$$\left\{\frac{1-(1-x)^{1/2}}{1+(1-x)^{1/2}}\right\}^{\lambda/2} \cong \left(\frac{x^{1/2}}{2}\right)^\lambda.$$

Also

$$(1-x)^{-1/4} \cong 1$$

and

$$\exp\{\lambda(1-x)^{1/2}\} \cong e^\lambda.$$

Thus, as $x \to 0$ for fixed λ,

$$x^{-1/2}w_1(x) \cong 2^{1/2}\left(\frac{x^{1/2}}{2}\right)^\lambda e^\lambda. \qquad (30)$$

Relations (27) through (30) now imply that

$$c_1(\lambda) = 2^{1/2}\lambda^{-\lambda}e^\lambda\Gamma(\lambda+1)$$

and

$$c_2(\lambda) = 0.$$

The solution $w_1(x)$ is thus a multiple of $x^{1/2}J_\lambda(\lambda x^{1/2})$ and, by the above

$$J_\lambda(\lambda x^{1/2}) = \frac{\lambda^\lambda e^{-\lambda}}{(2x)^{1/2}\Gamma(\lambda+1)}\,w_1(x).$$

One further simplification can be made. As $\lambda \to \infty$, *Stirling's formula* gives as an approximation to $\Gamma(\lambda+1)$

$$\Gamma(\lambda+1) \cong (2\pi\lambda)^{1/2}\lambda^\lambda e^{-\lambda}.$$

Thus, using this relation in (30), the behavior of $J_\lambda(\lambda x^{1/2})$ as $\lambda \to \infty$ is found to be

$$J_\lambda(\lambda x^{1/2}) \cong \tfrac{1}{2}(\pi\lambda x)^{-1/2} w_1(x). \tag{31}$$

Using (26) and (31), we thus have that as $\lambda \to \infty$ with x in I'

$$J_\lambda(\lambda x^{1/2}) = (2\pi\lambda)^{-1/2}(1-x)^{-1/4}\left\{\frac{1-(1-x)^{1/2}}{1+(1-x)^{1/2}}\right\}^{\lambda/2}$$

$$\cdot \exp\{\lambda(1-x)^{1/2}\}\left\{1 + O\left(\frac{1}{\lambda}\right)\right\}.$$

Note that although relation (31) was derived using approximations valid in I', the constants $c_1(\lambda)$ and $c_2(\lambda)$ in (27) do not depend on x. Relation (31) must thus hold for all real x. In particular, suppose we use our general method to find approximations to $w_1(x)$ valid in intervals I'' where $p_0(x)$ is negative. Relation (31) then immediately gives approximations to $J_\lambda(\lambda x^{1/2})$ valid as $\lambda \to \infty$ for x in I''.

2.5   Comparison Equations and Constructions

In the last two sections, the asymptotic behavior of the solutions of a differential equation was obtained by a straightforward method. In the remainder of this chapter, a vastly more powerful technique will be introduced. With proper modification, this technique not only handles subintervals I' of I on which $p_0(x)$ does not vanish, but also easily handles the turning point case. For simplicity, the differential equation

$$w'' - \lambda^2 p(x)w = 0 \tag{32}$$

will be treated first.

The heart of this technique involves a transformation and the use of a *comparison equation*. Roughly speaking, a comparison equation for (32) is an equation of the form

$$y'' - \lambda^2 q(x)y = 0 \tag{33}$$

whose solutions are known and where $q(x)$ "models" the behavior of $p(x)$ on a particular subinterval I'. For example, on a subinterval I' on which $p(x)$ is strictly positive, we might take $q(x) = +1$. Then, $q(x)$ is also strictly positive on I' and a comparison equation for (32) is

$$y'' - \lambda^2 y = 0$$

having the known solutions $\exp\{+\lambda x\}$ and $\exp\{-\lambda x\}$. If $p(x)$ is strictly negative on I', we might choose $q(x) = -1$ and use as a comparison equation

$$y'' + \lambda^2 y = 0.$$

The real solutions of this equation are $\sin(\lambda x)$ and $\cos(\lambda x)$. On the other hand, if $p(x)$ has a simple zero at a point x_0 in I', a relevant choice is $q(x) = x - x_0$ leading to the comparison equation,

$$y'' - \lambda^2(x - x_0)y = 0.$$

Solutions of this equation are $Ai[\lambda^{2/3}(x - x_0)]$ and $Bi[\lambda^{2/3}(x - x_0)]$. Thus, in each of the above cases, the solutions of the comparison equation are known and the function $q(x)$ models the behavior of $p(x)$ on I'.

To use a comparison equation in obtaining approximations to the solutions of Equation (32), we must first find a transformation which transforms our original equation into the relevant comparison equation plus "something else." This transformation will usually involve the independent variable as well as the dependent variable. To help keep track of things, the transformed variables will be denoted by Greek letters.

Since we are working with two independent variables, it is important to avoid ambiguity in the use of a prime to denote differentiation. To do this, we will adopt the convention that a prime always denotes differentiation with respect to the argument of the differentiated function. For example, if w is a function of x and ϕ is a function of η, then $w' = dw/dx$ but $\phi' = d\phi/d\eta$.

Suppose that for the subinterval I' under consideration, a relevant comparison equation is,

$$\phi'' - \lambda^2 \gamma(\eta)\phi = 0 \qquad (34)$$

where ϕ and γ are functions of a new independent variable η, as yet undetermined, and $\gamma(\eta)$ has the same behavior on the transformed interval I'' as $p(x)$ has on I'. Thus, $\gamma(\eta)$ plays the same role as $q(x)$. The solutions of Equation (34) are assumed known and will be denoted by $\bar{\phi}(\eta)$. We need a transformation

$$\begin{aligned} \eta &= f(x), \\ \phi(\eta) &= g(x)w(x), \end{aligned} \qquad (35)$$

under which the original Equation (32) transforms to the equation

$$\phi'' - \lambda^2\{\gamma(\eta) + \delta(\eta, \lambda)\}\phi = 0. \qquad (36)$$

The function $\delta(\eta, \lambda)$ is thus the "something else" mentioned previously. It might be expected that if $\delta(\eta, \lambda)$ is small as $\lambda \to \infty$ compared to $\gamma(\eta)$, then first approximations to the solutions $\phi(\eta)$ of the transformed equation (36) as $\lambda \to \infty$ would be given by the known solutions $\bar{\phi}(\eta)$ of the comparison equation (34). This is in fact the case. Now, knowing first approximations to solutions of the transformed equation, first approximations to the solutions of our original Equation (32) are easily obtained by use of the transformation (35). We find that

$$w(x) = \frac{1}{g(x)} \phi(\eta) \approx \frac{1}{g(x)} \tilde{\phi}[f(x)] \tag{37}$$

as $\lambda \to \infty$. Thus, the key to this technique is deciding on the relevant comparison equation for the particular subinterval under consideration and determining the functions $f(x)$ and $g(x)$ in the transformation (35).

So far we have considered the solutions of the comparison equation (34) as providing only a first approximation to the solutions of the transformed equation (36). However, higher approximations to solutions of Equation (36) may also be obtained from solutions of the comparison equation. This is done by means of a *construction* of the form

$$\phi(\eta) = \alpha(\eta, \lambda)\tilde{\phi}(\eta) + \frac{1}{\lambda^2}\beta(\eta, \lambda)\tilde{\phi}'(\eta). \tag{38}$$

In this construction, $\alpha(\eta, \lambda)$ and $\beta(\eta, \lambda)$ are assumed to have expansions of the form

$$\alpha(\eta, \lambda) = \sum_{n=0}^{\infty} \frac{\alpha_n(\eta)}{\lambda^{2n}}$$

and

$$\beta(\eta, \lambda) = \sum_{n=0}^{\infty} \frac{\beta_n(\eta)}{\lambda^{2n}}.$$

The problem of obtaining higher approximations to the solutions $\phi(\eta)$ of the transformed equation as $\lambda \to \infty$ is thus reduced to the problem of obtaining the coefficients $\alpha_n(\eta)$ and $\beta_n(\eta)$ in the construction. A scheme for doing this will be derived in Section 2.8. However, it is clear that since solutions of $\tilde{\phi}(\eta)$ of the comparison equation provide first approximations to solutions $\phi(\eta)$ of the transformed equation, we must have

$$\alpha_0(\eta) = 1.$$

Again, once higher approximations to $\phi(\eta)$ have been obtained, higher approximations to the solutions $w(x)$ of our original equation may be easily obtained through the transformation (35).

2.6   The Liouville-Green Transformation

The transformation-comparison equation technique will now be used to obtain first approximations to solutions of Equation (32) on subintervals I' on which $p(x)$ does not vanish. The function $p(x)$ will be assumed strictly positive on I'. The case where $p(x)$ is strictly negative is left as an exercise.

Since $p(x)$ is strictly positive on I', we choose $\gamma(\eta) = 1$ and take as a comparison equation

$$\phi''(\eta) - \lambda^2 \phi(\eta) = 0. \tag{39}$$

The exact solutions of Equation (39) are

$$\tilde{\phi}_1(\eta) = e^{\lambda \eta}$$

and

$$\tilde{\phi}_2(\eta) = e^{-\lambda \eta}. \tag{40}$$

We must now determine the functions $f(x)$ and $g(x)$ in the transformation (35) which correspond to our choice of a comparison equation. This will be done in some detail.

Let

$$w(x) = \{\eta'(x)\}^s \phi(\eta), \tag{41}$$

where the transformed independent variable η is an as yet undetermined function of x and the exponent s is to be chosen so that the transformed equation has no first derivative term. By the chain rule for differentiation

$$\frac{dw}{dx} = \eta' \frac{d\phi}{d\eta}.$$

Thus

$$w'(x) = s\eta'^{s-1}\eta''\phi(\eta) + \eta'^{s+1}\phi'(\eta)$$

and

$$w''(x) = \{s(s-1)\eta'^{s-2}\eta''^2 + s\eta'^{s-1}\eta'''\}\phi(\eta)$$
$$+ \{(2s+1)\eta'^s\eta''\}\phi'(\eta) + \eta'^{s+2}\phi''(\eta).$$

It is obvious that if the transformed equation is to have no first derivative term, we should take

$$s = -\tfrac{1}{2}.$$

Then $w(x) = \{\eta'(x)\}^{-1/2}\phi(\eta)$,

$$w''(x) = \{\tfrac{3}{4}\eta'^{-5/2}\eta'' - \tfrac{1}{2}\eta'^{-3/2}\eta'''\}\phi(\eta) + \eta'^{3/2}\phi''(\eta),$$

and Equation (32) transforms to the equation

$$\phi'' - \lambda^2\{\gamma(\eta) + \delta(\eta, \lambda)\}\phi = 0$$

with

$$\gamma(\eta) = p(x)\{\eta'(x)\}^{-2} \tag{42}$$

and

$$\delta(\eta, \lambda) = -\frac{1}{\lambda^2}\left\{\frac{3}{4}\frac{\eta''^2}{\eta'^4} - \frac{1}{2}\frac{\eta'''}{\eta'^3}\right\}. \tag{43}$$

Since we have chosen $\gamma(\eta) = 1$, (42) implies

$$p(x)\{\eta'(x)\}^{-2} = 1$$

or

$$\eta = \int^x \{p(t)\}^{1/2}\, dt \qquad (44)$$

and hence

$$w(x) = \{p(x)\}^{-1/4}\phi(\eta). \qquad (45)$$

We have thus found the required functions $f(x)$ and $g(x)$ which correspond to our choice of (39) as a comparison equation. The transformation (44) and (45) is usually called the *Liouville-Green Transformation*.

The function $\delta(\eta, \lambda)$ can be put in a somewhat simpler form. Let

$$\psi(\eta) = \frac{\eta''}{\eta'^2}.$$

Then $\delta(\eta, \lambda)$ is expressible in terms of $\psi(\eta)$ and its first derivative. Since

$$\psi'(\eta) = \frac{\eta'''}{\eta'^3} - 2\psi^2(\eta),$$

we have

$$\delta(\eta, \lambda) = \frac{1}{4\lambda^2}\{\psi^2(\eta) + \tfrac{1}{4}\psi'(\eta)\}.$$

The extent to which solutions $\bar{\phi}(\eta)$ of the comparison equation provide good first approximations to solutions $\phi(\eta)$ of the transformed equation depends on the relative sizes of $\gamma(\eta)$ and $\delta(\eta, \lambda)$ as $\lambda \to \infty$. To explore this point, it is worthwhile to use a mixed notation and express $\delta(\eta, \lambda)$ in terms of $p(x)$ and its derivatives. Since

$$\eta'(x) = \{p(x)\}^{1/2}$$

and

$$\eta''(x) = \frac{1}{2}\frac{p'(x)}{\{p(x)\}^{1/2}},$$

we have

$$\psi(\eta) = \frac{1}{2}\frac{p'(x)}{\{p(x)\}^{3/2}}.$$

Similarly,

$$\frac{\eta'''}{\eta'^3} = \frac{1}{2\{p(x)\}^3}[p(x)p''(x) - \tfrac{1}{2}\{p(x)\}^2].$$

Thus

$$\delta(\eta, \lambda) = \frac{1}{\lambda^2} \left\{ \frac{4pp'' - 5p'^2}{16p^3} \right\}$$

$$= -\frac{1}{\lambda^2} p^{-3/4} \frac{d^2}{dx^2} (p^{-1/4}). \tag{46}$$

For $\delta(\eta, \lambda)$ to be small compared to $\gamma(\eta)$ as $\lambda \to \infty$, because of the factor λ^{-2} in (46), it is clearly sufficient that $\{p(x)\}^{-1/4}$ be a slowly varying, twice differentiable function of x on the interval I'. In most cases, this condition is satisfied. Note that if

$$p(x) = \{ax + b\}^{-4},$$

where a and b are arbitrary constants such that $p(x) > 0$ on I', then $\delta(\eta, \lambda) \equiv 0$ and our original equation is transformed identically into the comparison equation. It should also be noted that $p(x)$ may have double poles in I' since this behavior is cancelled out in $\delta(\eta, \lambda)$.

If, as $\lambda \to \infty$, $\delta(\eta, \lambda)$ is small compared to $\gamma(\eta) = 1$, then, using (40) and (44), good approximations to the solutions $\phi(\eta)$ of the transformed equation are given by

$$\phi(\eta) \approx e^{\pm \lambda \eta} = \exp\left\{ \pm \lambda \int^x (p(t))^{1/2} \, dt \right\}.$$

Using the Liouville-Green transformation now gives as first approximations to the solutions $w(x)$ of our original equation

$$w(x) \approx \{p(x)\}^{-1/4} \exp\left\{ \pm \lambda \int^x (p(t))^{1/2} \, dt \right\}. \tag{47}$$

Note that the right-hand side of (47) is identical to the function $\tilde{w}(x)$ obtained from the WKB method. The feature of the WKB results which is not duplicated in (47) is the vague error bound $\{1 + O(1/\lambda)\}$. Error bounds for first approximations obtained through the methods in this section will be derived in Section 2.8.

At first glance, it may seem that the transformation–comparison equation approach is unnecessarily complex compared to the more straightforward WKB method. After all, we must introduce a new independent variable and treat the error separately. However, this added complexity is more than compensated for by the fact that with slight modification we can deal with the turning point case which is not accessible using the WKB method. Also, the error bounds obtained are sharp and explicit rather than vague.

To illustrate the use of the Liouville-Green transformation, consider the Airy equation

$$w'' - \lambda^2 x w = 0. \tag{48}$$

Approximations to the solution $w_2(x)$ of (48), which is a multiple of $Ai(\lambda^{2/3}x)$, were derived in Section 2.5. Since $p(x) = x$, take as I' the set of bounded x strictly greater than zero so that $p(x)$ is strictly positive on I'. The Liouville-Green transformation in this case is

$$\eta = \tfrac{2}{3}x^{3/2} \qquad (x > 0)$$

and

$$w(x) = x^{-1/4}\phi(\eta).$$

Under this transformation, Equation (48) becomes

$$\phi'' - \lambda^2\left\{1 - \frac{5}{24\lambda^2\eta^2}\right\}\phi = 0, \tag{49}$$

and it is clear that for x in I'

$$\delta(\eta, \lambda) = -\frac{5}{24\lambda^2\eta^2}$$

will be small compared to $\gamma(\eta) = 1$. The exponentially small solution of the comparison equation

$$\phi'' - \lambda^2\phi = 0$$

will thus be a good first approximation to the desired solution of the transformed equation (49).

The arguments of this section may easily be extended to the more general case

$$w'' - \lambda^2 p(x, \lambda)w = 0, \tag{50}$$

where $p(x, \lambda)$ does not vanish on the interval I' under consideration. If $p(x, \lambda)$ is strictly positive on I', we may again use the comparison equation (39). The only modification necessary is to replace $p(x)$ by $p(x, \lambda)$. For example, the Liouville-Green transformation is now

$$\eta = \int^x \{p(t, \lambda)\}^{1/2}\, dt$$

and

$$w(x) = \{p(x, \lambda)\}^{-1/4}\phi(\eta),$$

which takes Equation (50) into the transformed equation

$$\phi'' - \lambda^2\{1 + \delta(\eta, \lambda)\}\phi = 0, \tag{51}$$

where

$$\delta(\eta, \lambda) = -\frac{1}{\lambda^2}\{p(x, \lambda)\}^{-3/4}\frac{d^2}{dx^2}\{p(x, \lambda)\}^{-1/4}. \tag{52}$$

Note that in the earlier case where $p(x, \lambda) = p(x)$, the expression for $\delta(\eta, \lambda)$ was separable, e.g., $\delta(x, \lambda) = \lambda^{-2}\delta(x)$. This will usually not happen in the more general case (52). Also, in the more general case, it is harder to judge the size of $\delta(\eta, \lambda)$. However, if $\delta(\eta, \lambda)$ is small compared to 1, the analogue of expression (47) is clearly

$$w(x) \approx \{p(x, \lambda)\}^{-1/4} \exp\left\{\pm \lambda \int^x \{p(t, \lambda)\}^{1/2} \, dt\right\}. \tag{53}$$

For purposes of deriving error bounds for the approximations given by (53), it is convenient to rewrite the transformed equation (51) in the form

$$\phi'' - \{\lambda^2 + \breve{\delta}(\eta, \lambda)\}\phi = 0,$$

where

$$\breve{\delta}(\eta, \lambda) = \lambda^2 \delta(\eta, \lambda).$$

In particular, in the case where $p(x, \lambda) = p(x)$, $\breve{\delta}(\eta, \lambda)$ is a function only of η.

2.7   Error Bounds

We will now obtain explicit error bounds for the first approximations of Section 2.6 obtained through use of the comparison equation (39) and the Liouville-Green transformation. It is clearly sufficient to obtain the error $\epsilon(\eta, \lambda)$ associated with the use of solutions $\breve{\phi}(\eta)$ of the comparison equation as first approximations to the solutions $\phi(\eta)$ of the transformed equation since if

$$\phi(\eta) = \breve{\phi}(\eta)\{1 + \epsilon(\eta, \lambda)\},$$

then

$$w(x) = \{p(x, \lambda)\}^{-1/4}\breve{\phi}(\eta)\{1 + \epsilon(\eta, \lambda)\} \, .$$

The following theorems are due to F.W.J. Olver:

Theorem A: *Let λ be a positive parameter and let $\breve{\delta}(\eta, \lambda)$ be a continuous function of η in the interval $a \leq \eta \leq b$. Then, in this interval the differential equation*

$$\phi'' - \{\lambda^2 + \breve{\delta}(\eta, \lambda)\}\phi = 0 \tag{54}$$

has solutions $\phi_1(\eta)$ and $\phi_2(\eta)$ such that

$$\phi_1(\eta) = e^{\lambda\eta}\{1 + \epsilon_1(\eta, \lambda)\} \tag{55}$$

and

$$\phi_2(\eta) = e^{-\lambda\eta}\{1 + \epsilon_2(\eta, \lambda)\}, \tag{56}$$

$$|\epsilon_k(\eta, \lambda)| \le \exp\left\{\frac{F_k(\eta, \lambda)}{2\lambda}\right\} - 1 \quad (k = 1, 2)$$

$$(57)$$

with

$$F_1(\eta, \lambda) = \int_a^\eta |\breve{\delta}(t, \lambda)| \, dt \qquad (58)$$

and

$$F_2(\eta, \lambda) = \int_\eta^b |\breve{\delta}(t, \lambda)| \, dt. \qquad (59)$$

The interval (a, b) may be infinite provided that the integrals in (58) and (59) converge.

Theorem B: *With the conditions of Theorem A, the differential equation*

$$\phi'' - \{-\lambda^2 + \breve{\delta}(\eta, \lambda)\}\phi = 0 \qquad (60)$$

has solutions $\phi_1(\eta)$ and $\phi_2(\eta)$ such that

$$\phi_1(\eta) = \sin(\lambda\eta) + \epsilon_1(\eta, \lambda) \qquad (61)$$

and

$$\phi_2(\eta) = \cos(\lambda\eta) + \epsilon_2(\gamma, \lambda) \qquad (62)$$

where

$$|\epsilon_k(\eta, \lambda)| \le \exp\left\{\frac{F(\eta, \lambda)}{\lambda}\right\} - 1 \quad (k = 1, 2)$$

with

$$F(\eta, \lambda) = \left|\int_c^\eta |\breve{\delta}(t, \lambda)| \, dt\right|, \qquad (63)$$

c being an arbitrary point such that $a \le c \le b$. The interval (a, b) and the value of c may be infinite provided the integral in (63) converges.

Proof of Theorem A

Let $\Delta(\eta, \lambda) = e^{\lambda\eta}\epsilon_1(\eta, \lambda)$ so that

$$\phi_1(\eta) = e^{\lambda\eta} + \Delta(\eta, \lambda).$$

Then, from Equation (54), $\Delta(\eta, \lambda)$ satisfies the differential equation

$$\Delta'' - \lambda^2 \Delta = \check{\delta}(\eta, \lambda)\{\Delta + e^{\lambda\eta}\}. \tag{64}$$

Since the homogeneous equation

$$\Delta'' - \lambda^2 \Delta = 0$$

has solutions $\exp\{\lambda\eta\}$ and $\exp\{-\lambda\eta\}$ and the Wronskian $W(\eta)$ of these two solutions is a constant, a particular solution of Equation (64) has the form [cf. relation (4)]

$$\Delta(\eta, \lambda) = \frac{1}{2\lambda} \int_a^\eta \{e^{\lambda(\eta-t)} - e^{\lambda(t-\eta)}\}$$
$$\cdot \check{\delta}(t, \lambda)\{\Delta(t, \lambda) + e^{\lambda t}\} \, dt, \quad (65)$$

which is an integral equation for $\Delta(\eta, \lambda)$. Equation (65) will be solved by the method of *successive approximations*. We introduce a sequence of iterates $\{\Delta_n(\eta, \lambda)\}$ defined by

$$\Delta_0(\eta, \lambda) = 0$$

and, for $n \geq 1$,

$$\Delta_n(\eta, \lambda) = \frac{1}{2\lambda} \int_a^\eta \{e^{\lambda(\eta-t)} - e^{\lambda(t-\eta)}\}\check{\delta}(t, \lambda) \cdot \{\Delta_{n-1}(t, \lambda) + e^{\lambda t}\} \, dt. \tag{66}$$

Since the integration variable t is less than η in (66),

$$0 \leq e^{\lambda\eta} - e^{\lambda(2t-\eta)} \leq e^{\lambda\eta}. \tag{67}$$

For $n = 1$,

$$\Delta_1(\eta, \lambda) = \frac{1}{2\lambda} \int_a^\eta \{e^{\lambda\eta} - e^{\lambda(2t-\eta)}\}\check{\delta}(t, \lambda) \, dt,$$

and thus, using (67)

$$|\Delta_1(\eta, \lambda)| \leq \frac{e^{\lambda\eta}}{2\lambda} \int_a^\eta |\check{\delta}(t, \lambda)| \, dt \tag{68}$$

$$= \frac{e^{\lambda\eta} F_1(\eta, \lambda)}{2\lambda},$$

and similarly,

$$\Delta_2(\eta, \lambda) - \Delta_1(\eta, \lambda) = \frac{1}{2\lambda} \int_a^\eta \{e^{\lambda(\eta-t)} - e^{\lambda(t-\eta)}\}\check{\delta}(t, \lambda) \, \Delta_1(t, \lambda) \, dt$$

$$= \frac{1}{2\lambda} \int_a^\eta \{e^{\lambda\eta} - e^{\lambda(2t-\eta)}\}\check{\delta}(t, \lambda)e^{-\lambda t} \, \Delta_1(t, \lambda) \, dt.$$

From (67) and (68), we thus have that

$$|\Delta_2(\eta, \lambda) - \Delta_1(\eta, \lambda)| \leq \frac{e^{\lambda\eta}}{2\lambda} \int_a^\eta |\delta(t, \lambda)| \frac{F_1(t, \lambda)}{2\lambda} \, dt.$$

Since

$$|\delta(t, \lambda)| = \frac{dF_1(t, \lambda)}{dt}, \quad F_1(t, \lambda) \frac{dF_1(t, \lambda)}{dt} = \frac{d}{dt} \left\{ \frac{F_1^2(t, \lambda)}{2} \right\}$$

and

$$F_1(a, \lambda) = 0,$$

we find

$$|\Delta_2(\eta, \lambda) - \Delta_1(\eta, \lambda)| \leq \frac{e^{\lambda\eta}}{2} \left\{ \frac{F_1(\eta, \lambda)}{2\lambda} \right\}^2.$$

Similarly, by induction, it can be shown that for $n \geq 0$

$$|\Delta_{n+1}(\eta, \lambda) - \Delta_n(\eta, \lambda)| \leq \frac{e^{\lambda\eta}}{(n+1)!} \left\{ \frac{F_1(\eta, \lambda)}{2\lambda} \right\}^{n+1} \qquad (69)$$

We wish to show that the sequence of iterates $\{\Delta_n(\eta, \lambda)\}$ converges as $n \to \infty$ to a limit function $\Delta(\eta, \lambda)$ which is a solution of the integral equation (65). To do this, note that $\Delta_{n+1}(\eta, \lambda)$ can be written as the collapsing series

$$\Delta_{n+1}(\eta, \lambda) = \sum_{j=0}^n \{\Delta_{j+1}(\eta, \lambda) - \Delta_j(\eta, \lambda)\}. \qquad (70)$$

The convergence of the sequence $\{\Delta_n\}$ to a limit function $\Delta(\eta, \lambda)$ as $n \to \infty$ is thus equivalent to the convergence of the series in (70) and, if the series converges, we will have

$$\Delta(\eta, \lambda) = \sum_{j=0}^\infty \{\Delta_{j+1}(\eta, \lambda) - \Delta_j(\eta, \lambda)\} \qquad (71)$$

However, by (69), each term in the above series is less in absolute value than $e^{\lambda\eta}$ times a term in the convergent series for

$$\exp\left\{ \frac{F_1(\eta, \lambda)}{2\lambda} \right\}.$$

Thus, the series in (71) is absolutely convergent and hence converges uniformly in $a \leq \eta \leq b$. The sequence of iterates

$\{\Delta_n(\eta, \lambda)\}$ thus converges to a limit function $\Delta(\eta, \lambda)$ and, from (69) and (71),

$$|\Delta(\eta, \lambda)| \le \sum_{j=0}^{\infty} |\Delta_{j+1}(\eta, \lambda) - \Delta_j(\eta, \lambda)|$$

$$\le \sum_{j=0}^{\infty} \frac{e^{\lambda\eta}}{(j+1)!} \left\{ \frac{F_1(\eta, \lambda)}{2\lambda} \right\}^{j+1}.$$

Summing the above series we obtain

$$|\Delta(\eta, \lambda)| \le e^{\lambda\eta} \left\{ \exp\left[\frac{F_1(\eta, \lambda)}{2\lambda} \right] - 1 \right\}.$$

Although it has not been rigorously proved, $\Delta(\eta, \lambda)$ is clearly a solution of the integral equation (65). Since $\epsilon_1(\eta, \lambda) = e^{-\lambda\eta} \Delta(\eta, \lambda)$, relation (57) of Theorem A is established for $k = 1$. The proof of relation (57) for the case $k = 2$ is similar and will be omitted. The proof of Theorem B also follows the same lines and is left as a problem at the end of this chapter.

Several remarks on Theorem A might be made at this point. From (57), (58), and (59), it is clear that

$$\epsilon_1(a, \lambda) = \epsilon_2(b, \lambda) = 0.$$

Thus, if a is finite, we have obtained a first approximation to a solution $\phi_1(\eta)$ of (54) which satisfies the boundary condition

$$\phi_1(a) = e^{\lambda a}.$$

Similarly, if $a = -\infty$, then the solution $\phi_1(\eta)$ of (54) is such that

$$\phi_1(\eta) \cong e^{\lambda\eta} \quad \text{as} \quad \eta \to -\infty.$$

Similar remarks hold for $\phi_2(\eta)$ as $\eta \to b$.

In (58) and (59), $F_1(\eta, \lambda)$ and $F_2(\eta, \lambda)$ are defined in terms of $\delta(\eta, \lambda)$ with the integration being over an η-interval. Suppose that on an interval where $p(x, \lambda)$ is strictly positive, we have transformed the equation

$$w'' - \lambda^2 p(x, \lambda)w = 0$$

into Equation (54) by use of a Liouville-Green transformation. In this case, it is clearly more convenient to employ a mixed notation and define $F_1(\eta, \lambda)$ and $F_2(\eta, \lambda)$ in terms of $p(x, \lambda)$ and its derivatives with the integration being over an x-interval. Consider, for example, $F_1(\eta, \lambda)$ and let the point $\eta = a$ correspond to the point $x = \tilde{a}$. Then, using the expression (52) for $\delta(\eta, \lambda)$ and the fact that in (58)

$$\frac{d}{ds} t(s) = \eta'(s) = \{p(s)\}^{1/2},$$

$F_1(\eta, \lambda)$ can be written in the form

$$F_1(\eta, \lambda) = \int_a^x \{p(s, \lambda)\}^{-1/4} \left| \frac{d^2}{ds^2} \{p(s, \lambda)\}^{-1/4} \right| ds. \qquad (72)$$

A similar result holds for $F_2(\eta, \lambda)$.

As an example of Theorem A, error bounds will be derived for the solution $w_2(x)$ of Equation (48) which is a multiple of $\mathrm{Ai}(\lambda^{2/3}x)$. In this case,

$$p(x, \lambda) = x.$$

Clearly,

$$w_2(x) = x^{-1/4}\phi_2(\eta)$$

is the desired solution. By Theorem A,

$$w_2(x) = x^{-1/4} \exp\{-\tfrac{2}{3}\lambda x^{3/2}\}\{1 + \epsilon_2(\eta, \lambda)\}.$$

Let $\eta = b$ correspond to $x = \bar{b}$. Then, since

$$\eta = \tfrac{2}{3}x^{3/2},$$

taking $b = +\infty$ in (59) corresponds to taking $\bar{b} = +\infty$. Thus, using an expression like (72) for $F_2(\eta, \lambda)$, we find

$$F_2(\lambda, \lambda) = \int_x^\infty s^{-1/4} \left| \frac{d^2}{ds^2} (s^{-1/4}) \right| ds$$

$$= \int_x^\infty \tfrac{5}{16} s^{-5/2}\, ds = \tfrac{5}{24} x^{-3/2}$$

and hence

$$|\epsilon_2(\eta, \lambda)| \le \exp\left\{\frac{5x^{-3/2}}{48\lambda}\right\} - 1.$$

This is much more explicit than the vague bound

$$\epsilon_2(\eta, \lambda) = O\left(\frac{1}{\lambda}\right)$$

obtained from the WKB method.

2.8 Higher Approximations

Before going on to the turning point case in Section 2.9, a scheme will be developed for obtaining higher approximations to the solutions $\phi(\eta)$ of the equation

$$\phi'' - \{\lambda^2 \gamma(\eta) + \delta(\eta, \lambda)\}\phi = 0$$

from solutions $\breve{\phi}(\eta)$ of the comparison equation

$$\phi'' - \lambda^2 \gamma(\eta)\phi = 0$$

where $\gamma(\eta) = \pm 1$. Only the case $\gamma(\eta) = +1$ will be considered.

As mentioned in Section 2.5, the basis of the scheme is a construction of the form

$$\phi(\eta) = \alpha(\eta, \lambda)\breve{\phi}(\eta) + \frac{1}{\lambda^2}\beta(\eta, \lambda)\breve{\phi}'(\eta) \qquad (73)$$

where $\alpha(\eta, \lambda)$ and $\beta(\eta, \lambda)$ are assumed to have expansions of the form

$$\alpha(\eta, \lambda) \cong \sum_{n=0}^{\infty} \frac{\alpha_n(\eta)}{\lambda^{2n}}$$

and

$$\beta(\eta, \lambda) \cong \sum_{n=0}^{\infty} \frac{\beta_n(\eta)}{\lambda^{2n}}$$

with $\alpha_n(\eta)$ and $\beta_n(\eta)$ both $O(1)$ for all n and

$$\alpha_0(\eta) = 1.$$

To obtain the relations satisfied by $\alpha(\eta, \lambda)$ and $\beta(\eta, \lambda)$ we first differentiate the construction (73) twice using the fact that

$$\breve{\phi}'' = \lambda^2 \breve{\phi}.$$

In particular,

$$\phi'' = \{\lambda^2\alpha + \alpha'' + 2\beta'\}\breve{\phi} + \left\{2\alpha' + \beta + \frac{1}{\lambda^2}\beta''\right\}\breve{\phi}'. \qquad (74)$$

Substituting (73) and (74) into the full differential equation and equating the coefficients of $\breve{\phi}(\eta)$ and $\breve{\phi}'(\eta)$ to zero yields the two coupled equations

$$\alpha''(\eta, \lambda) - \alpha(\eta, \lambda)\breve{\delta}(\eta, \lambda) + 2\beta'(\eta, \lambda) = 0,$$

$$2\alpha'(\eta, \lambda) + \frac{1}{\lambda^2}\{\beta''(\eta, \lambda) - \beta(\eta, \lambda)\breve{\delta}(\eta, \lambda)\} = 0.$$

Substituting the expansions for $\alpha(\eta, \lambda)$ and $\beta(\eta, \lambda)$ into these equations and equating powers of λ^{-2} to zero gives sets of two coupled differential equations (one set for each power of λ^{-2}) for the functions $\alpha_n(\eta)$ and $\beta_n(\eta)$. Each set has the form

$$\alpha_n''(\eta) - \breve{\delta}(\eta, \lambda)\alpha_n(\eta) + 2\beta_n'(\eta) = 0 \qquad (75)$$

$$2\alpha_{n+1}'(\eta) + \beta_n''(\eta) - \breve{\delta}(\eta, \lambda)\beta_n(\eta) = 0 \qquad (76)$$

with $n \geq 0$.

Since $\alpha_0(\eta)$ is known, consider Equation (75) to be an equation for $\beta_n(\eta)$. Then, by direct integration

$$\beta_n(\eta) = \frac{1}{2} \int^{\eta} \{\delta(t, \lambda)\alpha_n(t) - \alpha_n''(t)\} \, dt. \tag{77}$$

Having obtained $\beta_n(\eta)$, Equation (76) can be integrated directly to give

$$\alpha_{n+1}(\eta) = -\frac{1}{2} \beta_n'(\eta) + \frac{1}{2} \int^{\eta} \delta(t, \lambda)\beta_n(t) \, dt. \tag{78}$$

It is now clear that $\alpha_n(\eta)$ and $\beta_n(\eta)$ can be determined successively using (77) and (78). For example, taking $n = 0$ in (77), we can determine $\beta_0(\eta)$. Now, taking $n = 0$ in (78) $\alpha_1(\eta)$ can be determined. This process can be continued and thus, in principle, using (77) and (78) and the construction (73), approximations to solutions of the equation

$$\phi'' - \{\lambda^2 + \delta(\eta, \lambda)\}\phi = 0$$

can be determined to within any desired error as $\lambda \to \infty$. A similar scheme can be developed for using the construction (73) to obtain approximations to solutions in the case $\gamma(\eta) = -1$.

One further point should be noted. Suppose that v is a nonzero constant of order one, and

$$\delta(\eta, \lambda) = v + \bar{\delta}(\eta, \lambda).$$

Then, from (77), taking $n = 0$ and η_0 as the lower limit of integration,

$$\beta_0(\eta) = \frac{1}{2} v(\eta - \eta_0) + \frac{1}{2} \int_{\eta_0}^{\eta} \bar{\delta}(t, \lambda) \, dt.$$

Since $\beta_0(\eta)$ cannot be infinite, if $\eta_0 = -\infty$ some modifications must be made.

Define a new parameter μ by

$$\mu = (\lambda^2 + v)^{1/2}$$

and write the original equation in the form

$$\phi'' - \{\mu^2 + \bar{\delta}(\eta, \lambda)\}\phi = 0.$$

Now, use the equation

$$\phi'' - \mu^2\phi = 0$$

as a new comparison equation. It is clear that since $\bar{\delta}(\eta, \lambda)$ contains no constant term we avoid the previous difficulty. Also, the new comparison equation more closely resembles the original equation since the constant term from $\delta(\eta, \lambda)$ has been included. We thus obtain better approximations at each stage. For this reason, redefinition of the large parameter is advisable, even if the lower limit of integration η_0 is finite.

2.9 The Turning Point Case

The behavior of solutions of the differential equation

$$w'' - \lambda^2 p(x)w = 0 \tag{79}$$

will now be considered in the neighborhood of a point x_0 of I where $p(x)$ has a simple zero, i.e., $p(x_0) = 0$ but $p'(x_0) \neq 0$. For convenience, we will take $x_0 = 0$ and assume that $p(x) > 0$ for $x > 0$. These choices cause no restrictions. For example, if x_0 is not zero we can always make the preliminary translation $\tilde{x} = x - x_0$.

The transformation–comparison equation technique will again be used. To avoid confusion with the earlier case, the transformed independent variable will be denoted by ζ. However, the transformed dependent variable will still be denoted by ϕ.

As a comparison equation for (79) in the neighborhood of the point $x = 0$ we take the equation

$$\phi'' - \lambda^2 \xi \phi = 0. \tag{80}$$

Thus, as in Equation (79), the coefficient of the undifferentiated term in (80) has a simple zero at the origin. Exact solutions of Equation (80) are the Airy functions. Denoting these solutions by $\tilde{\phi}(\xi)$ and letting

$$\zeta = \lambda^{2/3}\xi,$$

we thus have

$$\tilde{\phi}_1(\xi) = Bi(\zeta) \tag{81}$$

and

$$\tilde{\phi}_2(\xi) = Ai(\zeta). \tag{82}$$

Relation (81) is sometimes called a *stretching transformation* and ζ is called a *stretched variable*. This is because when $\xi = O(\lambda^{-2/3})$, by (81), $\zeta = O(1)$. Thus, an interval about $\xi = 0$ on the ξ-axis of length $O(\lambda^{-2/3})$ is "stretched" by (81) to an interval about $\zeta = 0$ on the ζ-axis of length $O(1)$.

We must now obtain a transformation which takes our original equation (79) into a transformed equation of the form

$$\phi'' - \{\lambda^2 \xi - \delta(\xi)\}\phi = 0. \tag{83}$$

Clearly, if $\delta(\xi)$ is small compared to $\lambda^2 \xi$, then the solutions $\tilde{\phi}_1(\xi)$ and $\tilde{\phi}_2(\xi)$ of the comparison equation will be good first approximations to solutions of the transformed equation (83). To find the required transformation, we let

$$w(x) = \{\xi'(x)\}^{-1/2}\phi(\xi),$$

where ξ is an as yet undetermined function of x. Then, just as in Section 2.6, Equation (79) transforms into the equation

$$\phi'' - \{\lambda^2 \gamma(\xi) + \delta(\xi)\}\phi = 0 \qquad (84)$$

with

$$\gamma(\xi) = p(x)\{\xi'(x)\}^{-2}$$

and

$$\delta(\xi) = -\left\{\frac{p(x)}{\xi(x)}\right\}^{-3/4} \frac{d^2}{dx^2}\left\{\frac{p(x)}{\xi(x)}\right\}^{-1/4}.$$

Comparing (83) and (84) gives

$$p(x)\{\xi'(x)\}^{-2} = \xi(x)$$

or

$$\{\xi(x)\}^{1/2}\xi'(x) = \{p(x)\}^{1/2}.$$

The required transformation is thus

$$w(x) = \left\{\frac{p(x)}{\xi(x)}\right\}^{-1/4} \phi(\xi) \qquad (85)$$

with

$$\xi(x) = \left[\frac{3}{2}\int_0^x \{p(t)\}^{1/2}\, dt\right]^{2/3} \quad (x \ge 0),$$

$$\xi(x) = -\left[\frac{3}{2}\int_x^0 \{-p(t)\}^{1/2}\, dt\right]^{2/3} \quad (x \le 0). \qquad (86)$$

Relations (85) and (86) are called a *Langer Transformation*. Note that neither $\xi'(x)$ nor $\delta(\xi)$ vanish at the turning point $x = 0$. In fact

$$\delta(0) = \frac{10p'(0)p'''(0) - 9\{p''(0)\}^2}{140\{p'(0)\}^{8/3}}.$$

If $\delta(\xi)$ is small compared to $\lambda^2\xi$, then using relation (85) from the Langer transformation and the solutions in relation (82), good first approximations to two linearly independent solutions of Equation (79) are easily obtained. We find

$$w_1(x) \approx \left\{\frac{p(x)}{\xi(x)}\right\}^{-1/4} Bi(\zeta) \qquad (87)$$

and

$$w_2(x) \approx \left\{\frac{p(x)}{\xi(x)}\right\}^{-1/4} Ai(\zeta). \qquad (88)$$

Both of these approximations are well defined at $x = 0$. In fact, expanding $p(x)$ in a power series about $x = 0$ gives

$$\zeta(x) = \{p'(0)\}^{1/3} x + O(x^2),$$

and thus

$$\left\{ \frac{p(x)}{\zeta(x)} \right\}^{-1/4}_{x=0} = \{p'(0)\}^{-1/6}.$$

Since

$$Ai(0) = \frac{3^{-2/3}}{\Gamma(2/3)},$$

$$Bi(0) = \frac{3^{-1/6}}{\Gamma(2/3)},$$

we find that

$$w_1(0) = \frac{3^{-2/3}\{p'(0)\}^{-1/6}}{\Gamma(2/3)}$$

and

$$w_2(0) = \frac{\{3p'(0)\}^{-1/6}}{\Gamma(2/3)}.$$

These values are useful to have since they provide starting values for direct numerical integration of Equation (79).

As was the case with approximations coming from the Liouville-Green transformation, a separate discussion is needed to obtain error bounds for the first approximations (87) and (88). Such bounds may be obtained by the methods of Section 2.7. However, the proofs of analogous theorems are more difficult due to the more complex nature of the Airy functions. The construction (73) of Section 2.8 may again be used to obtain higher approximations to the solutions of Equation (83) from the solutions of the comparison equation (80). A scheme for obtaining the functions $\alpha_n(\zeta)$ and $\beta_n(\zeta)$ is developed in the problems at the end of this chapter.

2.10 The Connection Problem

The behavior of solutions of Equation (79) has now been obtained both on subintervals I' of I on which $p(x)$ does not vanish and in the neighborhood of a simple turning point where $p(x)$ has a simple zero. If we again suppose the turning point is at $x = 0$, that $p(x)$ has no other zeros on I, and that $p(x) > 0$ for $x > 0$, $p(x) < 0$ for $x < 0$, then we have the following situation on I:

(a) For $x < 0$, approximations to $w(x)$ are expressible in terms of sines and cosines and have oscillatory behavior.

(b) In the neighborhood of $x = 0$, approximations to $w(x)$ are expressible in terms of Airy functions having as arguments a stretched variable.

(c) For $x > 0$, approximations to $w(x)$ are expressible in terms of growing and decaying exponentials.

In particular, suppose that we want approximations to a solution $w(x)$ of Equation (79) which is exponentially small when $p(x) > 0$. Then, $w(x)$ must be a multiple of the solution $w_2(x)$. In the Liouville-Green transformation, take the unspecified lower limit of integration to be the turning point $x = 0$. Using a mixed notation, we then have

$$w(x) \approx c_2(\lambda)\{+p(x)\}^{-1/4} \exp\{-\lambda\eta\},$$

for $x > 0$ where $\eta' = \{p(x)\}^{1/2}$,

$$w(x) \approx b(\lambda)\left\{\frac{p(x)}{\xi(x)}\right\}^{-1/4} Ai(\zeta)$$

in a neighborhood of $x = 0$ with $\zeta = \lambda^{2/3}\xi$ and $\xi = \xi(x)$ as in (86), and

$$w(x) \approx c_2(\lambda) \cdot c\{-p(x)\}^{-1/4} \cos\{\lambda\eta + a\}$$

for $x < 0$. The constant $b(\lambda)$ is used to *normalize* the approximations. It is an arbitrary constant and, in a particular problem, may be given any suitable value. The constants $c_2(\lambda)$, c, and a are not arbitrary, but, for the moment, are undetermined.

The approximation containing the Airy function is called an *inner approximation*. The approximations involving the cosine and exponential are called *outer approximations*. In three regions there are thus three different approximations to the same function $w_2(x)$. We are thus faced with the problem of smoothly connecting the different approximations.

This connection is made possible by the fact that the domains of validity of both outer approximations overlap the domain of validity of the inner approximation. Consider, for example, the connection of the inner approximation with the outer approximation for $x > 0$. The situation is shown in Figure 2.2. In Region A, $\lambda^{2/3}x$ is $O(1)$ while

	$x = 0$		
Region A		Region B	Region C
$x \ll 1$		$x \ll 1$	$x = O(1)$
$\lambda^{2/3}x = O(1)$		$\lambda^{2/3}x \gg 1$	

Figure 2.2

$x << 1$. In Region B, we still have $x << 1$ but now $\lambda^{2/3} x >> 1$. On the other hand, in Region C, x is $O(1)$.

The inner approximation is valid in Regions A and B, while the exponential type outer approximation is valid in Regions B and C. Both types of approximations are thus valid in Region B and can be connected by requiring that in this overlapping region they match, i.e., are asymptotically equivalent. In the course of this matching, the proper choice of the constant $c_2(\lambda)$ will be determined.

To accomplish this matching, note first that

$$\eta(x) = \tfrac{2}{3}\{\xi(x)\}^{3/2}. \tag{89}$$

Also, in Region B, the stretched variable ζ is large and we may use the asymptotic expansion for $Ai(\zeta) = Ai(\lambda^{2/3}\xi)$ in the inner approximation. From a previous example, we have

$$Ai(\zeta) \approx \tfrac{1}{2}\pi^{-1/2}\lambda^{-1/6}\xi^{-1/4} \exp\{-\tfrac{2}{3}\lambda\xi^{3/2}\}.$$

The matching condition in Region B is thus

$$c_2(\lambda)\{p(x)\}^{-1/4} \exp\{-\lambda\eta\} =$$
$$b(\lambda)\left(\frac{p(x)}{\xi(x)}\right)^{-1/4} \tfrac{1}{2}\pi^{-1/2}\lambda^{-1/6}\{\xi(x)\}^{-1/4} \exp\{-\tfrac{2}{3}\lambda\xi^{3/2}\}.$$

Thus, taking

$$c_2(\lambda) = \tfrac{1}{2}\pi^{-1/2}\lambda^{-1/6}b(\lambda)$$

and using (89), it is clear that the two types of approximations asymptotically match in Region B and we have the desired connection. A similar connection can be made between the inner approximation and the oscillatory type outer approximation for $x < 0$. The correct choices for the constants a and c are

$$c = 2e^{-\pi i/4} \qquad \text{and} \qquad a = -\pi/4.$$

Thus, the oscillatory and exponentially decaying approximations are smoothly connected through the inner approximation.

So far only simple zeros of $p(x)$ have been considered. However, the generalization to points x_0 where $p(x)$ has a zero of order n is clear. The comparison equation for (79) should be such that the coefficient of the ϕ term also has a zero of order n, e.g.

$$\phi'' - \lambda^2 \xi^n \phi = 0,$$

where

$$w(x) = \{\xi'(x)\}^{-1/2}\phi(\xi)$$

and

$$p(x)\{\xi'(x)\}^{-2} = \{\xi(x)\}^n.$$

In the case $n = 2$, the solutions of the comparison equation are parabolic cylinder functions.

Problems

1. Construct approximations to two linearly independent solutions of Equation (24) for $x > 1$.

2. Use the transformation–comparison equation technique to obtain first approximations to the solutions of the equation

$$w'' - \lambda^2(x^2 - 1)w = 0$$

on the interval $0 < x < 2$. Obtain error bounds for the outer approximations. Show that the different approximations can be smoothly connected.

3. Prove Theorem B using the fact that if u is real then

$$|\sin(u)| \leq 1 \quad \text{and} \quad |\cos(u)| \leq 1.$$

4. Show that a less precise bound for $\epsilon_1(\eta, \lambda)$ in Theorem A is given by

$$|\epsilon_1(\eta, \lambda)| \leq \frac{\tfrac{1}{2}F_1(\eta, \lambda)}{\lambda - \tfrac{1}{2}F_1(\eta, \lambda)}.$$

(*Hint*: $(n + 1)! \geq 2^n$.)

5. If $p(x)$ has a simple zero at a point x_0, show that the Langer transformation takes the equation

$$w'' - \{\lambda^2 p(x) + q(x)\}w = 0$$

into the equation

$$\phi'' - \left\{\lambda^2 \xi + \delta(\xi) + \frac{\xi(x)q(x)}{p(x)}\right\}\phi = 0.$$

6. In obtaining higher approximations to solutions $\phi(\xi)$ of the equation

$$\phi'' - \{\lambda^2 \xi + \delta(\xi, \lambda)\}\phi = 0$$

from solutions $\bar{\phi}(\xi)$ of the comparison equation

$$\phi'' - \lambda^2 \xi \phi = 0$$

a construction of the form

$$\phi(\xi) = \alpha(\xi, \lambda)\bar{\phi}(\xi) + \frac{1}{\lambda^2}\beta(\xi, \lambda)\bar{\phi}'(\xi)$$

may be used. Show that the second derivative of this constructed solution is of the form

$$\phi''(\xi) = \{\lambda^2 \xi \alpha + \alpha'' + \beta + 2\xi\beta'\}\bar{\phi}(\xi)$$

$$+ \left\{2\alpha' + \xi\beta + \frac{1}{\lambda^2}\beta''\right\}\bar{\phi}'(\xi).$$

Use this expression to show that $\alpha(\xi, \lambda)$ and $\beta(\xi, \lambda)$ satisfy the coupled differential equations

$$\alpha'' - \delta\alpha + 2\xi\beta' + \beta = 0$$

and

$$2\alpha' + \frac{1}{\lambda^2}\{\beta'' - \delta\beta\} = 0.$$

Show that the functions $\alpha_n(\xi)$ and $\beta_n(\xi)$ in the expansions for $\alpha(\xi, \lambda)$ and $\beta(\xi, \lambda)$ satisfy coupled sets of differential equations of the form

$$\alpha_n'' - \delta\alpha_n + 2\xi\beta_n' + \beta_n = 0$$

and

$$2\alpha_{n+1}' + \beta_n'' - \delta\beta_n = 0$$

and hence

$$\beta_n(\xi) = \tfrac{1}{2}\xi^{-1/2}\int^\xi t^{-1/2}\{\delta(t, \lambda)\alpha_n(t) - \alpha_n''(t)\}dt$$

and

$$\alpha_{n+1}(\xi) = -\tfrac{1}{2}\beta_n'(\xi) = \tfrac{1}{2}\int^\xi \delta(t, \lambda)\beta_n(t)dt.$$

References

1. M. Abramowitz and I. A. Stegun, *Handbook of Mathematical Functions*, National Bureau of Standards, Washington, D.C., 1965.
2. E. A. Coddington, *An Introduction to Ordinary Differential Equations*, Prentice-Hall, Englewood Cliffs, N. J., 1965.
3. A. Erdelyi, *Asymptotic Expansions*, Dover, New York, 1956.
4. F. W. J. Olver, *Phil. Trans. A*, 247 (1954), 307–368.
5. F. W. J. Olver, *Proc. Cam. Phil. Soc.*, 57 (1961), 790–810.
6. L. Fraenkel, *Proc. Camb. Phil. Soc.*, 65 (1969) 209–284.
7. J. Heading, *An Introduction to Phase Integral Methods*, Methuen, London, 1962.

3

Singular Perturbations—
The Method of Matched
Asymptotic Expansions

3.1 Regular and Singular Perturbations

Let

$$D\{w(x); \epsilon\} = 0 \qquad (1)$$

denote a linear, second order ordinary differential equation containing
a real parameter ϵ and having real valued solutions $w(x)$. The parameter
ϵ will be assumed strictly positive but very small. We wish to solve the
boundary value problem consisting of Equation (1) and a set of two
boundary conditions

$$B\{w(a), w(b); \epsilon\} = 0 \qquad (2)$$

which the solutions $w(x)$ of (1) must satisfy at the endpoints a and b
of the interval I under consideration. We will suppose that we can
directly solve the problem

$$D\{w(x); 0\} = 0, \qquad (3)$$

$$B\{w(a), w(b); 0\} = 0 \qquad (4)$$

obtained by formally setting $\epsilon = 0$ in (1) and (2).

By a *perturbation expansion* for $w(x)$, we will mean the asymptotic parameter expansion

$$w(x) \cong \sum_{n=0}^{\infty} w_n(x)\epsilon^n \qquad (5)$$

as $\epsilon \to 0$. Using the above expansion in (1) and (2) gives an equivalent sequence of boundary value problems, not involving ϵ, for the functions $w_n(x)$. The first such problem for the first approximation $w_0(x)$ has the form

$$D\{w_0(x); 0\} = 0$$

$$B\{w_0(a), w_0(b); 0\} = 0.$$

This problem was assumed solvable. Higher approximations $w_n(x)$ $(n \geq 1)$ are solutions of differential equations of the form

$$D\{w_n(x); 0\} = \Phi_n(x),$$

which satisfy appropriate boundary conditions. The function $\Phi_n(x)$ involves the known functions $w_0(x), w_1(x), \ldots, w_{n-1}(x)$.

As an example of how a perturbation scheme works, consider the differential equation

$$w'' - \epsilon w = 0$$

on the interval $0 \leq x \leq 1$. The boundary conditions to be satisfied are

$$w(0) - A = 0$$

and

$$w(1) - B = 0.$$

Using the perturbation expansion (5) for $w(x)$ gives

$$w_0''(x) + \sum_{n=1}^{\infty} \{w_n''(x) - w_{n-1}(x)\}\epsilon^n = 0,$$

$$\{w_0(0) - A\} + \sum_{n=1}^{\infty} w_n(0)\epsilon^n = 0,$$

and

$$\{w_0(1) - B\} + \sum_{n=1}^{\infty} w_n(1)\epsilon^n = 0.$$

Equating the coefficients of powers of ϵ to zero, we thus obtain an equivalent sequence of boundary value problems. The first approximation $w_0(x)$ is a solution of the problem

$$D\{w_0(x); 0\} = w_0''(x) = 0,$$

$$w_0(0) = A, \quad \text{and} \quad w_0(1) = B,$$

i.e.

$$w_0(x) = (B - A)x + A.$$

For $n \geq 1$, higher approximations $w_n(x)$ are solutions of the problems

$$D\{w_n(x), 0\} = w_n'' = w_{n-1},$$

$$w_n(0) = w_n(1) = 0.$$

In particular,

$$w_1''(x) = w_0(x) = (B - A)x + A,$$

$$w_1(0) = w_1(1) = 0,$$

so

$$w_1(x) = \frac{(B - A)}{3} x^3 + \frac{A}{2} x^2 - \frac{(A + 2B)}{6} x.$$

In general, for $n \geq 1$, $w_n(x)$ will be a polynomial of degree $2n + 1$ having no constant term. Note that this implies $w_n(x) = O(1)$ for all $0 \leq x \leq 1$. Since $\{\epsilon^n\}$ is an asymptotic sequence as $\epsilon \to 0$, the sequence $\{w_n(x)\epsilon_n\}$ will thus be an asymptotic sequence for all $0 \leq x \leq 1$ as $\epsilon \to 0$. This, in turn, implies that the expansion (5) is a valid asymptotic representation of the solution $w(x)$ as $\epsilon \to 0$ for *all* x in the interval I being considered. When this is the case, the perturbation is said to be *regular*. The method used in this example is thus a regular perturbation scheme.

A regular perturbation scheme is relatively simple and straightforward. Unfortunately, in many applications, there will be one or more subintervals I' of I on which the condition that $w_n(x) = O(1)$ for all n is not satisfied. For example, for $x \in I'$, $w_0(x)$ might be $O(1)$, but $w_1(x)$ is large like $1/\epsilon$. In this case, for $x \in I'$,

$$w_1(x)\epsilon = O(1) \neq o\{w_0(x)\},$$

and thus the expansion (5) cannot be a valid asymptotic representation for the solution $w(x)$ for $x \in I'$ as $\epsilon \to 0$. When this happens, the perturbation is said to be *singular*. The perturbation expansion (5) still remains valid outside of I' where $w_n(x) = O(1)$ for all n, but a new expansion is needed on I'.

One clue to singular behavior is the reduction in order of the differential equation (1) by the regular perturbation scheme. In the scheme, the first approximation $w_0(x)$ satisfies the differential equation (3) obtained by formally setting $\epsilon = 0$ in (1). If in (1) the coefficient of the second derivative term is ϵ, then Equation (3) can be of at most first order. For example, if Equation (1) has the form

$$\epsilon w'' + w' + w = 0, \tag{6}$$

then $w_0(x)$ is a solution of the equation

$$w_0'(x) + w_0(x) = 0. \tag{7}$$

At first glance, such reduction in order may seem desirable since exact solutions of first order equations are easily obtained. If c is a constant, the general solution of Equation (7) is

$$w_0(x) = ce^{-x}.$$

However, the solution of Equation (7) must also satisfy two boundary conditions, for example

$$w_0(0) = A, \qquad w_0(1) = B.$$

It is extremely fortuitous if the solution of a first order equation can be made to satisfy two given conditions since we have only one arbitrary constant c at our disposal. In the present example, if A and B do not satisfy a special relation, $w_0(x)$ can be made to satisfy either the condition at $x = 0$ or the condition at $x = 1$, but not both conditions. In fact, without further work, it is not clear which of the two boundary conditions should be applied. What is clear, however, is that unless we are extremely lucky there will be a singular region where the expansion (5) is not valid, probably in the neighborhood of the boundary point where we cannot satisfy the given condition.

The situation can be even more drastic. For example, if we have the boundary value problem

$$\epsilon w'' - w = 0,$$

$$w(0) - A = 0, \qquad \text{and} \qquad w(1) - B = 0,$$

with A and B nonzero, then the first approximation in the perturbation scheme is a solution of the problem

$$w_0(x) = 0,$$

$$w_0(0) = A, \qquad \text{and} \qquad w_0(1) = B.$$

Since A and B are not zero, we cannot satisfy either boundary condition for $w_0(x)$. In fact, we will see later that in this problem there are two singular regions and the perturbation expansion (5), while valid in the interior of the interval, is not valid in the neighborhood of either boundary point.

It might be suspected from the preceding examples that a sufficient condition for a perturbation to be regular is that Equation (3) be of the same order as Equation (1). This is in fact the case and, if this condition is satisfied, one can then proceed in a straightforward manner. For this reason, only the singular cases will be considered in the remainder of this chapter.

There are many methods for obtaining the behavior of $w(x)$ in regions where the perturbation expansion (5) breaks down. This chapter is not meant to be exhaustive and only the method of matched asymptotic expansions will be considered. The basic ideas in this method have already been encountered in Chapter 2. They include a stretching transformation which focuses attention on the small singular regions and the asymptotic matching of the expansions valid in these regions with the expansion (5) valid outside such regions. Since it is difficult to write down a step-by-step procedure, the methodology will be illustrated by means of examples.

3.2   A Problem with a Known Solution

Let a, b, A, and B be constants. To illustrate the method of matched asymptotic expansions, we will study the boundary value problem

$$\epsilon w'' + aw' + bw = 0, \tag{8}$$

$$w(0) = A, \quad \text{and} \quad w(1) = B. \tag{9}$$

The case $a > 0$ will be treated first. In addition, we will assume that A and B are not both zero and that

$$A \neq Be^{b/a}.$$

This problem has been chosen for several reasons. First, although quite simple, the problem requires the use of rather general techniques. Also, if $f(x)$ is a function strictly greater than zero on $[0, 1]$, the methodology developed for (8) and (9) can immediately be generalized to deal with the equation

$$\epsilon w'' + f(x)w' + g(x)w = 0.$$

Last, and perhaps most important, an exact solution for (8) and (9) is easily obtained. Let r_1 and r_2 be roots of the indicial equation

$$\epsilon r^2 + ar + b = 0$$

associated with (8), i.e.

$$r_1 = \frac{1}{2\epsilon}\{-a + (a^2 - 4\epsilon b)^{1/2}\},$$

$$r_2 = \frac{1}{2\epsilon}\{-a - (a^2 - 4\epsilon b)^{1/2}\}.$$

For later use, we note that

$$r_1 = -\frac{b}{a} + O(\epsilon) \tag{10}$$

and

$$r_2 = -\frac{a}{\epsilon} + O(1) \tag{11}$$

as $\epsilon \to 0$. Equation (8) then has solutions $e^{r_1 x}$ and $e^{r_2 x}$. The exact solution of our boundary value problem is thus that linear combination of $e^{r_1 x}$ and $e^{r_2 x}$ satisfying the conditions in (9), i.e.

$$w(x) = \frac{1}{e^{r_1} - e^{r_2}} \{(B - Ae^{r_2})e^{r_1 x} + (Ae^{r_1} - B)e^{r_2 x}\}. \qquad (12)$$

The advantage of having an exact solution while attempting to develop general methods is obvious. For one thing, every time something in the method goes wrong, we can go back to (12) and see how we should proceed. More important, however, we can also use (12) to find out what went wrong. The insights developed in this way are invaluable when dealing with a general problem whose exact solution is unknown.

The first step in any perturbation scheme is to assume that the solution $w(x)$ has a perturbation expansion like (5). Substituting this expansion into (8) and (9), we then obtain a sequence of problems, not involving ϵ, for the approximations, $w_n(x)$. The first approximation $w_0(x)$ is a solution of the problem

$$aw_0' + bw_0 = 0,$$
$$w_0(0) = A, \quad \text{and} \quad w_0(1) = B. \qquad (13)$$

For $n \geq 1$, higher approximations $w_n(x)$ are solutions of the problems

$$aw_n' + bw_n = -w_{n-1}'',$$
$$w_n(0) = w_n(1) = 0. \qquad (14)$$

Let us first consider $w_0(x)$. If d_0 is an arbitrary constant, then $w_0(x)$ must be of the form

$$w_0(x) = d_0 \exp\left\{-\frac{b}{a} x\right\} \qquad (15)$$

and we are immediately in trouble. We have only one arbitrary constant in (15) yet $w_0(x)$ must satisfy two given boundary conditions. This is impossible unless either $A = B = 0$ or $A = Be^{b/a}$, and both of these cases have been excluded. Thus, only one of the two given conditions for $w_0(x)$ can be satisfied. We have a similar situation for each of the higher approximations. At each stage, one additional constant of integration enters and it must be chosen so as to satisfy two boundary conditions. For example, $w_1(x)$ is a solution of the equation

$$aw_1' + bw_1 = -\frac{d_0 b^2}{a^2} \exp\left\{-\frac{b}{a} x\right\}$$

and therefore, if d_1 is a second arbitrary constant of integration, $w_1(x)$ must be of the form

$$w_1(x) = \left\{d_1 - \frac{d_0 b^2}{a^3} x\right\} \exp\left\{-\frac{b}{a} x\right\}.$$

Again, both of the given boundary conditions cannot be satisfied by any choice of the single constant d_1. We are thus led to the conclusion that the perturbation expansion (5) for $w(x)$ cannot be valid on the entire interval, and there must be at least one singular region. Note that once the singular regions are located, we automatically know which, if any, of the two boundary conditions in (13) and (14) should be applied. If we find a single singular region, near $x = 0$ say, then (5) is valid at $x = 1$ so we should require that $w_0(1) = B$ and, for $n \geq 1$, $w_n(1) = 0$. If we find a singular region near both endpoints, then neither of the two conditions in (13) and (14) are appropriate.

To locate the singular regions and see what went wrong, let us go back to the exact solution (12) of this problem. First, note that since a has been assumed strictly positive, $-a/\epsilon$ is large and negative. Using (11), this implies

$$e^{r_2} = O(\epsilon)$$

as $\epsilon \to 0$. On the other hand, from (10),

$$e^{r_1} = e^{-b/a}\{1 + O(\epsilon)\}$$

as $\epsilon \to 0$.

Let us first consider the portion of the interval on which $x = O(1)$. This is the entire interval excluding a thin region near $x = 0$. Since $x = O(1)$, as $\epsilon \to 0$

$$e^{r_1 x} = e^{-(b/a)x}\{1 + O(\epsilon)\}.$$

but

$$e^{r_2 x} = O(\epsilon).$$

Using this information in (12), we find that for $x = O(1)$,

$$w(x) = B \exp\left\{\frac{b}{a}(1 - x)\right\}\{1 + O(\epsilon)\} \tag{16}$$

as $\epsilon \to 0$.

To see if (5) is valid for $x = O(1)$, suppose that

$$w_0(1) = B.$$

Then, from (15)

$$w_0(x) = B \exp\left\{\frac{b}{a}(1 - x)\right\}, \tag{17}$$

which agrees with the result obtained from the exact solution for $x = O(1)$. Two conclusions now follow. First, the correct boundary conditions to be applied in (13) and (14) are the ones at $x = 1$. Also, the perturbation expansion (5) must be valid for $x = O(1)$, i.e., on the entire interval excluding a thin region near $x = 0$. To examine this thin region near $x = 0$, let us again go back to the exact solution and assume that $x = O(\epsilon)$.

If x is $O(\epsilon)$, then $-ax/\epsilon$ is $O(1)$, Thus, while e^{r_2} is still $O(\epsilon)$, $e^{r_2 x}$ is now $O(1)$ and

$$e^{r_2 x} = e^{-ax/\epsilon}\{1 + O(\epsilon)\}$$

as $\epsilon \to 0$. As a result, from (12), we get that for $x = O(\epsilon)$,

$$w(x) = \{Ae^{-ax/\epsilon} + Be^{b/a}(1 - e^{-ax/\epsilon})\}\{1 + O(\epsilon)\}. \tag{18}$$

Suppose that (5) were valid for $x = O(\epsilon)$. In this case, we could set $w_0(0) = A$. Then, from (15),

$$w_0(x) = A \exp\left\{-\frac{b}{a} x\right\},$$

which looks nothing like (18). The conclusion must be that there is indeed a singular region, of thickness $O(\epsilon)$ near $x = 0$, in which the perturbation expansion (5) breaks down and is not valid.

The exact behavior of the solution $w(x)$ of (8) and (9) is shown in Figure 3.1.

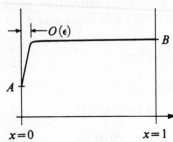

$$\epsilon w'' + aw' + bw = 0,$$
$$w(0) = A, \; w(1) = B,$$
$$a > 0$$

$O(\epsilon)$

B

A

$x = 0$ $x = 1$

Figure 3.1

We see that the behavior of $w(x)$ changes drastically as we go from $x = O(1)$ to $x = O(\epsilon)$. For $x = O(1)$, $w(x)$ is slowly varying. However, if $x = O(\epsilon)$, then small changes in x produce very large changes in $w(x)$. The reason for this is the rapidly varying exponential in (18). Such regions of rapid variation are called *boundary layers*. The boundary layer in this problem has a finite thickness which shrinks to zero as $\epsilon \to 0$. Because of this shrinking, it is convenient to say that the boundary layer occurs at a specified point, in this case at $x = 0$.

Why do we get a singular region at $x = 0$ in this problem? To answer this question, let us go back and take a closer look at what the perturbation expansion (5) does. Clearly, the effect of using (5) in (8) is to remove the second derivative term in the equation for $w_0(x)$ and to introduce only second derivatives of known functions in the equations for higher approximations. This is equivalent to the assumption that the term $\epsilon w''$ in (8) is small compared to the terms aw' and bw over the *entire* interval. To check this assumption, we will again use our exact solution (12). First, for $x = O(1)$,

$$\epsilon w''(x) = \epsilon \frac{Bb^2}{a^2} \exp\left\{\frac{b}{a}(1 - x)\right\}\{1 + O(\epsilon)\}$$

as $\epsilon \to 0$. This is indeed small compared to both aw' and bw, which are $O(1)$. However, for $x = O(\epsilon)$,

$$\epsilon w''(x) = \frac{a^2}{\epsilon}\{A - Be^{b/a}\}\exp\left\{-\frac{ax}{\epsilon}\right\}(1 + O(\epsilon)).$$

This term is of the same order as aw', which is also $O(1/\epsilon)$, and much larger than bw. As a result, in the boundary layer, the assumption which allows us to use the perturbation expansion (5) does not hold so (5) cannot be valid.

The importance of the term $\epsilon w''$ in the boundary layer cannot be overemphasized. In the next section, this fact provides the key to obtaining the boundary layer behavior of $w(x)$.

3.3 Boundary Layer Behavior

We will now obtain the behavior of the solution $w(x)$ of (8) and (9) in the boundary layer region. The constant a in (8) will still be assumed strictly positive. Also, the method developed does not use results obtained from the exact solution (12) and thus is quite general.

The method begins at the point where we have obtained $w_0(x)$ and recognized that since both boundary conditions in (13) cannot be satisfied there must be one or more boundary layers. Suppose that these boundary layers occur at the as yet undetermined points $\bar{x}$ in $[0, 1]$. To focus attention on these thin regions, let us define a stretching transformation

$$w(x) = \phi(\eta),$$
$$x - \bar{x} = \gamma(\epsilon)\eta,$$

(19)

where $\gamma(\epsilon)$ is as yet undetermined. The second derivative term in (8) is important in the boundary layer regions. Therefore, the transformation (19) must have the property that if an expansion of the form

$$\phi(\eta) = \sum_{n=0}^{\infty} \phi_n(\eta)\epsilon^n$$

(20)

is assumed for the solution of the transformed equation, each of the resulting problems contains a true second order equation. This condition will determine the function $\gamma(\epsilon)$.

Regardless of the value of $\bar{x}$, under the transformation (19), Equation (8) becomes

$$\phi'' + \alpha(\epsilon)\phi' + \beta(\epsilon)\phi = 0,$$

(21)

where

$$\alpha(\epsilon) = \frac{a\gamma(\epsilon)}{\epsilon}$$

and

$$\beta(\epsilon) = \frac{b\gamma^2(\epsilon)}{\epsilon}.$$

It is clear that if $\alpha(\epsilon)$ and $\beta(\epsilon)$ are at most $O(1)$, then the problems for $\phi_n(\eta)$ obtained by using the expansion (20) in Equation (21) will involve true second order equations. This suggests two possible choices for $\gamma(\epsilon)$. If $\gamma(\epsilon) = \epsilon^{1/2}$, then $\beta(\epsilon) = O(1)$. However for this choice, $\alpha(\epsilon) = a/\epsilon^{1/2} >> 1$ which is not allowable. We therefore choose

$$\gamma(\epsilon) = \epsilon \qquad (22)$$

so that $\alpha(\epsilon) = O(1)$. With this choice, $\beta(\epsilon) << 1$ and, in Equation (21), there is thus a balance between the ϕ'' and ϕ' terms. The relevant transformation and transformed equation are thus

$$w(x) = \phi(\eta),$$
$$x - \bar{x} = \epsilon\eta, \qquad (23)$$

and

$$\phi'' + a\phi' + \epsilon b\phi = 0. \qquad (24)$$

From (22), each of the boundary layers must have a width of $O(\epsilon)$. However, the location of the points $\bar{x}$ is still undetermined.

Since the expansion (20) involves a stretched variable η, it will be called an *inner expansion*. Similarly, the expansion (5) involving the original variable x will be called an *outer expansion*. The inner and outer expansions must, of course, be smoothly connected in some intermediate region where both expansions are valid. In this region, $x - \bar{x}$ is small while η is large.

Substituting the inner expansion into the transformed boundary layer equation (24), the first approximation $\phi_0(\eta)$ is found to be a solution of the second order equation

$$\phi_0'' + a\phi_0' = 0. \qquad (25)$$

Thus, if c_0 and c_1 are two arbitrary constants, then $\phi_0(\eta)$ is of the form

$$\phi_0(\eta) = c_0 - \frac{c_1}{a} e^{-a\eta}. \qquad (26)$$

Similarly, higher approximations $\phi_n(\eta)$ are solutions of the equations

$$\phi_n'' + a\phi_n' = -b\phi_{n-1}.$$

Using (26), we can now find the location of any boundary layers. If x lies *outside* the boundary layer, then, from (23), we must have $|\eta| \to \infty$ as $\epsilon \to 0$ since $|x - \bar{x}|$ is $O(1)$. Suppose there were a boundary layer at the right hand endpoint, i.e. $\bar{x} = 1$. Then, in the interior of the interval, $x - \bar{x}$ is negative and hence $\eta \to -\infty$ as $\epsilon \to 0$. Since $a > 0$, if $c_1 \neq 0$ then (26) becomes unbounded as we move from the boundary layer into the interior. However, we must be able to smoothly connect the inner and outer expansions since they are asymptotic representations for the same function. Clearly, this cannot be done if $\phi_0(\eta)$ becomes unbounded. Thus, if the boundary layer is at $\bar{x} = 1$, we must have $c_1 = 0$ so there is only one arbitrary constant c_0 in (26). In this case, we are again in trouble for an inner expansion valid at $x = 1$ must not only match to the outer expansion valid in the interior but must also satisfy the right hand boundary condition. With only one arbitrary constant at our disposal, both of these conditions cannot be satisfied. Therefore, there cannot be a boundary layer at the right hand endpoint.

A similar argument shows that a boundary layer in the interior of the interval is impossible. If this were not the case, then outside an interior boundary layer on the left we would have $\eta \to -\infty$ as $\epsilon \to 0$, and (26) again becomes unbounded unless $c_1 = 0$. Also, an inner expansion valid in an interior boundary layer must match the outer expansions on both the left and the right. Clearly, this matching is impossible with only the single constant c_0 at our disposal.

The only possible boundary layer is thus at the left hand endpoint. In this case, $\bar{x} = 0$ so $x - \bar{x}$ is positive. Thus, in the interior of the interval, $\eta \to +\infty$ as $\epsilon \to 0$. $\text{Exp}\{-a\eta\}$ is now well behaved and we retain both arbitrary constants in (26). As a result, we can both match inner and outer expansions and satisfy the boundary conditions at $x = 0$.

Having found that the boundary layer is at $x = 0$, the following statements can now be made.

(a) The inner expansion $\phi(\eta)$ is valid in a boundary layer of thickness $O(\epsilon)$ at $x = 0$. Since $x = 0$ implies $\eta = 0$, it must satisfy the boundary condition $\phi(0) = A$. Further, it must match to the outer expansion when $\eta \gg 1$.

(b) The outer expansion $w(x)$ is valid on the entire interval excluding a thin region near $x = 0$. It must satisfy the boundary condition $w(1) = B$ and match to the inner expansion for $x \ll 1$.

Let us now use this information to determine the three constants d_0, c_0, and c_1 in the first approximations (15) and (26). The boundary condition $w_0(1) = B$ immediately gives $d_0 = Be^{b/a}$ or

$$w_0(x) = B \exp\left\{\frac{b}{a}(1 - x)\right\}.$$

Also, for $x \ll 1$,

$$w_0(x) \cong Be^{b/a}. \tag{27}$$

The boundary condition $\phi_0(0) = A$ implies that

$$A = c_0 - \frac{c_1}{a}.$$

Also, for $\eta >> 1$,

$$\phi_0(\eta) \cong c_0. \tag{28}$$

From (27) and (28), the matching condition to lowest order is thus

$$c_0 = Be^{b/a}$$

which, in turn, implies

$$c_1 = a\{Be^{b/a} - A\}.$$

Thus, the first inner approximation is of the form

$$\phi_0(\eta) = Ae^{-a\eta} + Be^{b/a}\{1 - e^{-a\eta}\}. \tag{29}$$

Since

$$\eta = \frac{x}{\epsilon},$$

this result is identical to (18), which was obtained through use of the exact solution (12). However, note that nowhere in the derivation of (29) have we assumed any knowledge of an exact solution.

Up to this point, we have assumed that a was strictly positive. The analysis for the case where a is strictly negative is almost identical and will be omitted. As may be seen from (26), the effect of changing the sign of a is to move the boundary layer to the right hand endpoint $x = 1$. Figure 3.2 shows the behavior of the solution in this case.

$$\epsilon w'' + aw' + bw = 0,$$
$$w(0) = A, \ w(1) = B,$$
$$a < 0$$

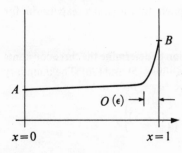

Figure 3.2

Let us now briefly consider the case where a is zero, and deal with the equation

$$\epsilon w'' + bw = 0. \tag{30}$$

For the moment, the sign of the constant b will not be specified. As is the case if a is nonzero, the first step in the perturbation scheme is to assume an outer expression like (5) for $w(x)$. The first approximation $w_0(x)$ is then a solution of the equation

$$bw_0(x) = 0.$$

Since b is not zero, $w_0(x)$ is then identically zero. In fact, we find that $w_n(x) = 0$ for all n. Suppose neither A nor B is zero so that we cannot satisfy either of the two boundary conditions in (9). Let us now define a stretching transformation, with $\gamma(\epsilon)$ unspecified, and look for boundary layers.

Applying the stretching transformation to Equation (30), we obtain the transformed Equation (21) with $\alpha(\epsilon) = 0$. The function $\gamma(\epsilon)$ should thus be chosen to make $\beta(\epsilon) = O(1)$, i.e.,

$$\gamma(\epsilon) = \epsilon^{1/2}. \tag{31}$$

This immediately tells us that any boundary layers will have a thickness of order $\epsilon^{1/2}$. With this choice for $\gamma(\epsilon)$, the transformed equation becomes

$$\phi'' + b\phi = 0. \tag{32}$$

Equation (32) is exactly solvable. However, we must now distinguish between the cases $b > 0$ and $b < 0$. The latter case will be considered first.

Suppose b is strictly negative. Then, if c_0 and c_1 are arbitrary constants, the general solution of (32) is

$$\phi(\eta) = c_0 \exp\{(-b)^{1/2}\eta\} + c_1 \exp\{-(-b)^{1/2}\eta\} \tag{33}$$

with

$$\eta = \frac{x - \bar{x}}{\epsilon^{1/2}}. \tag{34}$$

We can now determine at which points $\bar{x}$ in $[0, 1]$, there are boundary layers.

As in the case where a was nonzero, if x lies to the right of a boundary layer, then $\eta \to +\infty$ as $\epsilon \to 0$. If x lies to the left of a boundary layer, $\eta \to -\infty$ as $\epsilon \to 0$. Therefore, from (33) there can be no boundary layers in the interior of the interval.

Let us now examine the situation at the endpoint $x = 0$. The interior of the interval corresponds to $\eta \to +\infty$. Thus, for $\phi(\eta)$ to remain bounded, we must have $c_0 = 0$. Then $\phi(\eta) \to 0$ as $\eta \to +\infty$ and automatically matches to the identically zero outer expansion. The remaining constant c_1 can now be used to satisfy the boundary condition at $x = 0$. Thus, there is a boundary layer at $x = 0$ and

$$\phi(\eta) = A \exp\{-(-b)^{1/2}\eta\} \tag{35}$$

with

$$\eta = \frac{x}{\epsilon^{1/2}}. \tag{36}$$

The situation at the endpoint $x = 1$ is similar. Denote the stretched variable in this layer by $\tilde{\eta}$. The solution of (33) which satisfies the boundary condition at $x = 1$ and matches to the outer expansion is

$$\phi(\tilde{\eta}) = B \exp\{(-b)^{1/2}\tilde{\eta}\} \tag{37}$$

with

$$\tilde{\eta} = \frac{x - 1}{\epsilon^{1/2}}. \tag{38}$$

Exact behavior is shown in Figure 3.3a.

To complete the discussion of Equation (30), suppose that b is strictly positive. The general real solution of Equation (32) is then

$$\phi(\eta) = c_0 \sin(b^{1/2}\eta) + c_1 \cos(b^{1/2}\eta). \tag{39}$$

For no nonzero choices of the constants c_0 and c_1, does (39) remain near any fixed number as either $\eta \to +\infty$ or $\eta \to -\infty$. We thus have the curious situation where the outer expansion is identically zero yet there can be no thin singular regions. The only possible explanation is that the entire interval $[0, 1]$ must be a singular region and the constants c_0 and c_1 in (39) should be determined by the two boundary conditions in (9). The exact solution (12) confirms this explanation. It shows that the solution is wildly oscillatory over the entire interval with a period of $O(\epsilon^{1/2})$. A portion of this solution is shown in Figure 3.3b.

$$\epsilon w'' + bw = 0, \qquad w(0) = A, \qquad w(0) = B.$$

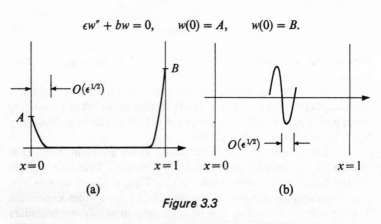

(a)   (b)

Figure 3.3

3.4   Composite Expansions

It is often convenient to use the matched inner and outer expansions to construct a single *composite expansion* valid on the entire interval I under consideration. This construction is made possible by the fact that the outer expansion and the inner expansions have terms in common, namely those terms which match in the intermediate regions where the stretched boundary layer variable is large. Additional terms in the outer expansion are asymptotically zero in the boundary layers. Similarly, additional terms in the inner expansions are asymptotically zero outside the boundary layers. As a result, if, using a mixed notation, we add the outer expansion and inner expansions and subtract from this sum the duplicated common terms, the expansion we obtain asymptotically represents the solution of the boundary value problem on the entire interval.

To illustrate this construction, consider the inner and outer expansions obtained in the case $a > 0$ in Section 3.3. The first approximations $w_0(x)$ and $\phi_0(\eta)$ are

$$w_0(x) = B \exp\left\{\frac{b}{a}(1 - x)\right\} \tag{40}$$

and

$$\phi_0(\eta) = Ae^{-a\eta} + Be^{b/a}\{1 - e^{-a\eta}\} \tag{41}$$

where

$$\eta = \frac{x}{\epsilon}.$$

The terms that (40) and (41) have in common are

$$B \exp\left\{\frac{b}{a}(1 - x)\right\} \quad \text{and} \quad Be^{b/a}, \text{respectively,}$$

since these are the terms which match for $x << 1$ but $\eta >> 1$. Additional terms in (41) are asymptotically zero outside the boundary layer since $\eta \to +\infty$ when $x = O(1)$.

The first approximation to the composite expansion $\psi(\eta, x)$ is thus

$$\psi_0(\eta, x) = Ae^{-a\eta} + Be^{b/a}\{e^{-(b/a)x} - e^{-a\eta}\}, \tag{42}$$

or, in terms of only the outer variable x,

$$\psi_0(x) = Ae^{-(a/\epsilon)x} + Be^{b/a}\{e^{-(b/a)x} - e^{-(a/\epsilon)x}\}.$$

Higher composite approximations $\psi_n(x)$ are constructed in a similar manner. In fact, if we let $L(x)$ and $M(\eta)$ denote those terms in the outer expansion $w(x)$ and inner expansion $\phi(x)$, respectively, which match in the intermediate region, then

$$\psi(x, \eta) = w(x) + \phi(\eta) - L(x)$$
$$= w(x) + \phi(\eta) - M(\eta).$$

For an example having more than one boundary layer, consider the case where $a = 0$ and $b < 0$. The outer expansion is

$$w(x) \equiv 0,$$

while the inner expansions are

$$\phi(\eta) = A \exp\{-(-b)^{1/2}\eta\},$$

and

$$\phi(\tilde{\eta}) = B \exp\{(-b)^{1/2}\tilde{\eta}\}$$

with

$$\eta = \frac{x}{\epsilon^{1/2}}$$

and

$$\tilde{\eta} = \frac{x - 1}{\epsilon^{1/2}}.$$

In terms of the outer variable x, the composite expansion in this case is thus

$$\psi(x) = A \exp\left\{-\left(\frac{-b}{\epsilon}\right)^{1/2} x\right\}$$

$$+ B \exp\left\{\left(\frac{-b}{\epsilon}\right)^{1/2}(x - 1)\right\}.$$

3.5   Linear Equations with Variable Coefficients

On the interval I, which for convenience will again be taken as $[0, 1]$, consider the boundary value problem

$$\epsilon w'' + f(x)w' + g(x)w = 0 \tag{43}$$

with

$$w(0) = A \quad \text{and} \quad w(1) = B. \tag{44}$$

The functions $f(x)$ and $g(x)$ will be assumed sufficiently differentiable on I. Unlike the constant coefficient case, an exact solution of this boundary value problem is generally not obtainable. As a result, in deriving the asymptotic behavior of the solution $w(x)$ of (43) and (44), we must often rely on "feel" or intuition. However, if $f(x)$ has constant sign on I, or if $f(x) \equiv 0$ but $g(x)$ has constant sign on I, the methods developed for the constant coefficient case are easily generalized. This generalization will be illustrated by considering the case where $f(x) > 0$ on I. If $f(x)$ has zeros at points in I, the situation is much more complicated and will be considered briefly at the end of this section.

Suppose that $f(x) > 0$ on I. This is the variable coefficient analogue of the case $a > 0$ in Sections 3.2 and 3.3. As always, our first step is to assume that $w(x)$ has an outer expansion of the form

$$w(x) = \sum_{n=0}^{\infty} w_n(x)\epsilon^n.$$

The first approximation $w_0(x)$ is then a solution of the boundary value problem

$$f(x)w_0' + g(x)w_0 = 0, \tag{45}$$

$$w_0(0) = A, \quad \text{and} \quad w_0(1) = B. \tag{46}$$

If c is an arbitrary constant, the general solution of Equation (45) is

$$w_0(x) = c \exp\left\{-\int_{\tilde{x}}^{x} \frac{g(t)}{f(t)} \, dt\right\}, \tag{47}$$

where the lower limit of integration $\tilde{x}$ in (47) is as yet undetermined. It is clear that $w_0(x)$ cannot be made to satisfy both boundary conditions in (46). We will thus need one or more boundary layers and inner expansions. Hence, the next step is to define a stretching transformation (19) which will explicitly bring out the importance of the second derivative term in the boundary layers. Applying the stretching transformation (19), Equation (43) becomes

$$\phi'' + \frac{\gamma(\epsilon)f(x)}{\epsilon} \phi' + \frac{\gamma^2(\epsilon)g(x)}{\epsilon} \phi = 0.$$

In the boundary layer, the correct balance in the above equation again involves the ϕ'' and ϕ' terms. This implies that $\gamma(\epsilon) = \epsilon$ or, if the boundary layer is at x_0,

$$x - x_0 = \epsilon\eta.$$

Expanding $f(x)$ and $g(x)$ in a power series about x_0, the boundary layer equation is thus

$$\phi'' + \{f(x_0) + O(\epsilon)\}\phi' + \epsilon\{g(x_0) + O(\epsilon)\}\phi = 0. \tag{48}$$

To determine the location of the boundary layers, assume an inner expansion of the form

$$\phi(\eta) = \sum_{n=0}^{\infty} \phi_n(\eta)\epsilon^n.$$

The first inner approximation is then a solution of the equation

$$\phi_0'' + f(x_0)\phi' = 0,$$

i.e., if c_0 and c_1 are constants

$$\phi_0(\eta) = c_0 - \frac{c_1}{f(x_0)} e^{-f(x_0)\eta}. \tag{49}$$

By the same arguments as in the constant coefficient case, since $f(x_0) > 0$, (49) implies that the boundary layer must be located at the left hand endpoint, and hence $x_0 = 0$. Thus, the relevant boundary condition for the first outer approximation $w_0(x)$ is $w_0(1) = B$. This condition will be satisfied if we take $c = B$ and $\bar{x} = 1$ in (47), i.e.

$$w_0(x) = B \exp\left\{\int_x^1 \frac{g(t)}{f(t)} \, dt\right\}. \tag{50}$$

The relevant boundary condition for the first inner approximation is $\phi_0(\eta) = A$, or, from (49),

$$c_0 - \frac{c_1}{f(x_0)} = A. \tag{51}$$

The condition that the inner and outer expansions should match in the intermediate region where $x << 1$ but $\eta >> 1$ gives, to lowest order, the additional relation

$$c_0 = B \exp\left\{\int_0^1 \frac{g(t)}{f(t)} \, dt\right\}. \tag{52}$$

The constant c_1 in (49) may now be found from (51), and we finally obtain

$$\phi_0(\eta) = A e^{-f(x_0)\eta}$$
$$+ B \exp\left\{\int_0^1 \frac{g(t)}{f(t)} \, dt\right\} \cdot \left\{1 - e^{-f(x_0)\eta}\right\}. \tag{53}$$

Higher approximations are derived in a similar manner.

If $f(x)$ has zeros at point $\bar{x}$ in I, it is clear from Equation (47) that unless $c = 0$ the outer expansion cannot be valid in a neighborhood of $\bar{x}$ since the integral in (47) would then be singular. This leads us to suspect that small neighborhoods of the points $\bar{x}$ may be singular regions. If $\bar{x}$ is an interior point of I, the singular region is called a *transition region*. Rather complex analysis is usually required in such cases.

Suppose that $f(x)$ has a simple zero at a single point $\bar{x}$ in I, i.e., $f(\bar{x}) = 0$ but $f'(\bar{x}) \neq 0$. The following general statements can be made.

(a)   If $0 < \bar{x} < 1$ and $f'(\bar{x}) > 0$, the outer expansion is valid outside a small neighborhood of $\bar{x}$ of width $O(\epsilon^{1/2})$. The cases $0 \leq x < \bar{x}$ and $\bar{x} > x \geq 1$ may be considered separately. Depending on the behavior of $g(x)$, the behavior of the solution $w(x)$ in the transition region at $\bar{x}$ will either be a rounded step, a cusp, a peak, or a corner.

(b) If $0 < \bar{x} < 1$ and $f'(\bar{x}) < 0$, the outer expansion is valid in the interior of the region. The correct choice of the constant in (47) is $c = 0$ so $w(x) \cong 0$. There are boundary layers at both endpoints.

(c) If $\bar{x} = 0$ but $f'(0) > 0$, there is a boundary layer at $x = 0$ of width $O(\epsilon^{1/2})$ and the outer expansion is valid outside this region. Similarly, if $\bar{x} = 1$ but $f'(1) > 0$, there is a boundary layer at $x = 1$.

An example of situation (a) is shown in Figure 3.4. Cases where $f(x)$ vanishes at more than one point in I, or has higher order zeros, are quite difficult and will be omitted.

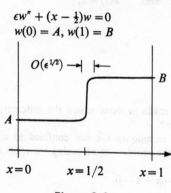

$$\epsilon w'' + (x - \tfrac{1}{2})w = 0$$
$$w(0) = A, \, w(1) = B$$

$O(\epsilon^{1/2})$

B

A

$x = 0 \qquad x = 1/2 \qquad x = 1$

Figure 3.4

If first approximations are adequate, cases where $f(x) \equiv 0$ and $g(x)$ vanishes at points in I can be most easily handled by the methods in Chapter 2. This is done by defining a new parameter

$$\lambda = \epsilon^{1/2}.$$

Clearly, $\lambda \to \infty$ as $\epsilon \to 0$, and we can then look for the asymptotic behavior of the solution $w(x)$ of the equation

$$w'' + \lambda^2 g(x)w = 0,$$

which satisfies the boundary conditions in (44).

Problems

1. In each of the following equations, the boundary conditions to be satisfied are

$$w(0) = 1 \quad \text{and} \quad w(1) = 2.$$

Find the location of the boundary layers and obtain first approximations to the outer, inner, and composite expansions.

a. $\epsilon w'' + 2w' = 0$
b. $\epsilon w'' + 2w' + 4w = 0$
c. $\epsilon w'' - 2w + 4w = 0$
d. $\epsilon w'' + 4w = 0$
e. $\epsilon w'' - 4w = 0$.

2. Obtain second approximations to the solution of the boundary value problem

$$\epsilon w'' + w' + w = 0,$$

$$w(0) = A_0 + \epsilon A_1, \quad \text{and} \quad w(1) = B_0 + \epsilon B_1.$$

3. Let c be a finite constant and consider the boundary value problem

$$\epsilon w'' + w' + w = 0,$$

$$w(0) = A, \quad \text{and} \quad w(x) \to c \quad \text{as } x \to \infty.$$

Show that the outer expansion breaks down for $x = O(1/\epsilon)$.

4. Obtain first approximations to the solutions of the boundary value problem

$$\epsilon w'' + f(x)w' + g(x)w = 0,$$

$$w(0) = 1, \quad \text{and} \quad w(1) = 2,$$

with

a. $f(x) = x + 2, g(x) = 1,$
b. $f(x) = x - 2, g(x) = 1,$
c. $f(x) = 0, g(x) = x - 2,$
d. $f(x) = \frac{1}{2} - x, g(x) = 0,$
e. $f(x) = x, g(x) = 1.$

5. What modifications must be made in cases where the differential equation (8) is inhomogeneous?

6. Applications of perturbation techniques are not confined to differential equations. Consider the problem of obtaining the roots of the polynomial

$$\epsilon x^3 - x + 1 = 0.$$

Show that assuming an expansion of the form

$$x = \sum_{n=0}^{\infty} x_n \epsilon^n$$

gives only one of the three roots. Why is this? Show that if r is an appropriately chosen rational number, the other two roots can be obtained through means of a stretching transformation

$$\epsilon^r x = y$$

and an inner expansion

$$y = \sum_{n=0}^{\infty} y_n \epsilon^{n/2}.$$

Find second approximations to all three roots.

Suggested References

1. J. D. Cole, *Perturbation Methods in Applied Mathematics*, Blaisdell, Waltham, Mass., 1968.
2. M. Van Dyke, *Perturbation Methods in Fluid Mechanics*, Academic Press, New York, 1964.
3. R. O'Malley, *Advances in Mathematics*, 2 (1968), 365–470.

4

The Existence of Periodic Solutions

4.1 Introduction

The question we now wish to consider is the following one: given a differential equation or system of differential equations, does there exist a periodic solution? If we think of the equation(s) as governing the behavior of some dynamical system, then we are asking whether there exists a periodically recurrent configuration or motion of the system.

Given that a periodic solution does exist, then a related question is the following one: If the initial data giving rise to the periodic solution is slightly perturbed, then what is the nature of the resulting solution? We can imagine two favorable cases, namely

(i)   one in which the new solution will also be periodic; or
(ii)   a stable situation in which the new solution tends toward the periodic solution as time proceeds.

A less favorable case would be the unstable situation where the resulting solution is no longer periodic and diverges away from the given periodic solution.

The two questions above have occupied a great deal of the research effort in ordinary differential equations in the last one hundred years. Furthermore much of the investigation deals with the second order differential equation. This might be expected inasmuch as in a dynamical model the solution can be thought of as a motion, in which case its second derivative corresponds to acceleration. The equating of the forces present will then usually lead to a second order equation. For this reason, and because we can often use the geometry of the plane to clarify the discussion, we will restrict ourselves to considering the second order ordinary differential equation.

Specifically let $f(t, x, y)$ be a real valued function defined for $-\infty < t < \infty$ and (x, y) in some region A of the xy-plane. We then consider the second order equation or its equivalent first order system

$$\text{(a)} \quad \ddot{x} = f(t, x, \dot{x}) \qquad \text{(b)} \quad \dot{x} = y,$$
$$\dot{y} = f(t, x, y). \tag{1}$$

If we assume that f, $\partial f/\partial x$, and $\partial f/\partial y$ are continuous then local existence and uniqueness of solutions is assured. Then the solution $x(t)$ of (1a) is a real valued twice continuously differentiable function, and the solution $(x(t), y(t))$ of (1b) will be a continuously differentiable pair. Furthermore from the first equation of (1b) we obtain the relation $\dot{x}(t) = y(t)$.*

Since we only wish to discuss periodic solutions it is natural in the case where f explicitly depends on t (the *nonautonomous* case) to require that f be periodic in t for every (x, y) in A. By this we mean there exists a positive number T such that

$$f(t + T, x, y) = f(t, x, y), \qquad -\infty < t < \infty, (x, y) \text{ in } A.$$

We will assume for simplicity that $T = 2\pi$ in the following discussion, so then a periodic solution (or 2π-periodic solution) is a function $x(t)$ satisfying

$$\ddot{x}(t) = f(t, x(t), \dot{x}(t)), \quad x(t + 2\pi) = x(t), \qquad -\infty < t < \infty.$$

If we note that all derivatives of a periodic function are periodic with the same period then it is clear why we require that f be periodic.

Examples

(i) $\ddot{x} + 2b\dot{x} + \omega^2 x = A \cos t$, where $b > 0$, ω and A are constants. The equation represents damped oscillatory motion with a periodic driving force or an RLC-circuit with an alternating

---

* In this chapter the variable t ($=$time) will be the independent variable, "dot" differentiation will be used, and x and y will be dependent variables. This is in keeping with hoary tradition and the physical significance of the problems to be discussed.

voltage source. In this case $f(t, x, \dot{x}) = -\omega^2 x - 2b\dot{x} + A \cos t$ and the corresponding first order system is

$$\dot{x} = y, \quad \dot{y} = -\omega^2 x - 2by + A \cos t.$$

(ii) $\ddot{x} + (\delta + \epsilon \cos t)x = 0$, where δ and ϵ are positive parameters. This is a form of the Mathieu equation which arises frequently in the study of stability of periodic solutions. Here $f(t, x, \dot{x}) = -(\delta + \epsilon \cos t)x$ and the equivalent first order system is

$$\dot{x} = y, \quad \dot{y} = -(\delta + \epsilon \cos t)x.$$

In the case where f does not explicitly depend on t (the *autonomous* case) then the equations (1) are of the form

(a) $\ddot{x} = g(x, \dot{x})$ (b) $\dot{x} = y,$

$$\dot{y} = g(x, y), \quad\quad (1')$$

where g is a real valued function continuous together with its first partial derivatives for (x, y) in A. In this case the period of any periodic solution is unspecified. We may regard the solutions $(x(t), y(t))$ of $(1'b)$ as parametric representations of continuously turning curves or trajectories $C: x = x(t), \quad y = y(t)$ in the xy-plane.

Another way of looking at this is to consider instead the first order differential equation

$$\frac{dy}{dx} = \frac{g(x, y)}{y}, \quad\quad (x, y) \text{ in } A.$$

A solution will be a curve C in the xy-plane which represents parametrically an infinite family of solutions of $(1'b)$. Since any periodic solution will satisfy $x(t) = x(t + T), y(t) = y(t + T), \quad -\infty < t < \infty$, for some $T > 0$, then the corresponding curve or trajectory C in the xy-plane will be a *closed curve*. Conversely any closed curve which is a solution of the previous first order equation represents a family of periodic solutions of $(1'b)$ and hence $(1'a)$.

Examples

(i) $\ddot{x} + \omega^2 x = 0$,   where ω is a constant. The equation represents simple undamped oscillatory motion. Here $g(x, \dot{x}) = -\omega^2 x$ and the equivalent first order system is

$$\dot{x} = y, \quad\quad \dot{y} = -\omega^2 x,$$

hence the corresponding first order equation is

$$\frac{dy}{dx} = \frac{-\omega^2 x}{y}.$$

Its variables are separable and solutions are found to be of the form $y^2 + \omega^2 x^2 = c^2$, c an arbitrary constant. These represent a family of ellipses in the xy-plane and we can conclude that all solutions of the original equation are periodic.

Each ellipse represents parametrically the infinite family of $2\pi/\omega$-periodic solutions

$$x(t) = \frac{c}{\omega} \cos(\omega t + \phi), \qquad y(t) = -c \sin(\omega t + \phi)$$

of the first order system, where ϕ is any arbitrary phase angle.

(ii) $\ddot{x} + \epsilon(x^2 - 1)\dot{x} + x = 0$, where ϵ is a positive parameter. This is the well studied Van der Pol equation which arises from certain electronic devices. We will show that for ϵ small, say $\epsilon < 0.1$, that a single isolated periodic solution exists.

This periodic solution has the property that nearby solutions tend toward it as t approaches infinity, hence it represents a stable configuration. Furthermore it can be approximated by $x(t) = 2 \cos(t + \phi)$, ϕ arbitrary, which is a solution of the equation when $\epsilon = 0$.

4.2 The Linear Equation

Perhaps the best place to begin is with an equation familiar to the reader, namely the second order linear equation with constant coefficients and a periodic driving term. One has the advantage of being able to describe solutions explicitly and furthermore some of the results have analogues in the case of nonlinear equations.

Two examples of physical systems which lead to these types of equations are the mass-spring system and the RLC-circuit diagrammed in Figure 4.1.

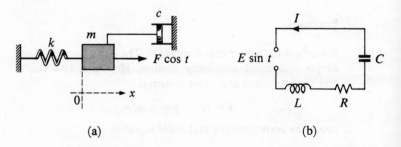

(a)   (b)

Figure 4.1

Both these systems lead to the equation

$$\ddot{x} + 2b\dot{x} + \omega^2 x = A \cos t, \qquad A, b > 0, \omega \neq 0, \tag{2}$$

where in (a)

x is the displacement, $b = c/2m$, $\omega^2 = k/m$, and $A = F/m$ with m = mass, k = spring constant, and c = coefficient of viscous damping,

and in (b)

x is the current I, $b = R/2L$, $\omega^2 = 1/LC$ and $A = E/L$ with R = resistance, L = inductance, and C = capacitance.

We will look for 2π-periodic solutions of (2).

The first case to examine is when $b = 0$ so no damping is present. Then depending on the value of ω the following occurs:

1. $\omega \neq$ integer: then solutions are of the form

$$x(t) = M \sin(\omega t + \phi) + \frac{A}{\omega^2 - 1} \cos t,$$

where M, ϕ are arbitrary constants. There is exactly one 2π-periodic solution, namely the solution

$$x(t) = \frac{A}{\omega^2 - 1} \cos t$$

obtained when $M = 0$. Another way of writing this solution is

$$x(t) = \frac{A}{|\omega^2 - 1|} \cos(t - \theta),$$

where θ is 0 or π depending on whether $\omega^2 - 1$ is positive or negative. This indicates whether the periodic solution is in or out of phase with the forcing term.

2. $\omega = n$, an integer, $n^2 \neq 1$: the solutions are

$$x(t) = M \sin(nt + \phi) + \frac{A}{n^2 - 1} \cos t.$$

Therefore all solutions are 2π-periodic.

3. $\omega^2 = 1$: this is the case of *resonance* and solutions are

$$x(t) = M \sin(t + \phi) + \frac{At}{2} \sin t,$$

so no 2π-periodic solutions exist. Here the solutions are oscillatory but become unbounded as t becomes infinite.

This completes the discussion when $b = 0$.

For the case $b \neq 0$ when damping is present the solutions are of the form

$$x(t) = X(t) + AG \cos(t - \theta),$$

where

$$G = [(\omega^2 - 1)^2 + 4b^2]^{-1/2}, \theta = \tan^{-1}2b/\omega^2 - 1, \qquad \text{and}$$

$$X(t) = \begin{cases} Me^{-bt} \sin(\omega_1 t + \phi), \omega_1 = \sqrt{\omega^2 - b^2}, & \text{if } b^2 - \omega^2 < 0, \\ e^{-bt}(Mt + N), & \text{if } b^2 - \omega^2 = 0, \\ Me^{-\lambda_1 t} + Ne^{-\lambda_2 t}, \quad \lambda_1, \lambda_2 > 0, & \text{if } b^2 - \omega^2 > 0, \end{cases}$$

where M, N and ϕ are arbitrary constants. Therefore there is only one periodic solution $x(t) = AG \cos(t - \theta)$ obtained when $X(t) \equiv 0$. Note in the case $b \neq 0$ that $X(t)$ decays exponentially to zero as t approaches $+\infty$, and for this reason $AG \cos(t - \theta)$ is called the *steady state solution*. The amplification factor G is a measure of the *gain* and θ is called the *steady state phase angle*.

The case $b \neq 0$, and case 1, when $b = 0$ are examples of the following more general result:

> Given the second order differential equation
>
> $$\ddot{x} + b(t)\dot{x} + c(t)x = f(t), \qquad (3)$$
>
> where $b(t)$, $c(t)$ and $f(t)$ are continuous and 2π-periodic, then it will have a 2π-periodic solution for any such $f(t)$ if and only if the homogeneous equation has no 2π-periodic solutions.

In the cases above $b(t)$ and $c(t)$ are constant, hence trivially 2π-periodic. A sketch of the proof follows; the reader should fill in the details.

First of all, (3) will have a 2π-periodic solution if and only if there exists a solution $x(t)$ satisfying the periodicity condition $x(0) = x(2\pi)$, $\dot{x}(0) = \dot{x}(2\pi)$. This follows from uniqueness of solutions and the fact that if $x(t)$ is a solution so is $x(t + 2\pi)$. Now let $x_1(t)$ and $x_2(t)$ be a fundamental pair of solutions of the homogeneous equation chosen so that

$$x_1(0) = 1, \quad \dot{x}_1(0) = 0; \quad x_2(0) = 0, \quad \dot{x}_2(0) = 1.$$

By the variation of parameters formula the general solution $x(t)$ satisfying $x(0) = \alpha$, $\dot{x}(0) = \beta$ is

$$x(t) = \alpha x_1(t) + \beta x_2(t) - x_1(t) \int_0^t \frac{f(s)x_2(s)}{W(s)}\, ds$$
$$+ x_2(t) \int_0^t \frac{f(s)x_1(s)}{W(s)}\, ds$$

where $W(s)$ is the Wronskian of $x_1(t)$ and $x_2(t)$. The periodicity condition then requires the solvability for α and β of equations of the following form:

$$(1 - x_1(2\pi))\alpha - x_2(2\pi)\beta = F_1,$$
$$-\dot{x}_1(2\pi)\alpha + (1 - \dot{x}_2(2\pi))\beta = F_2,$$

where the constants F_1 and F_2 depend on the choice of $f(t)$. Hence the determinant of coefficients must not vanish and this in turn will imply that if $\phi(t)$ is any solution of the homogeneous equation satisfying $\phi(0) = \phi(2\pi)$, $\dot{\phi}(0) = \dot{\phi}(2\pi)$, then it must be the trivial solution.

When $b = 0$ then cases 2 and 3 are examples of the following more general result.

> Given the second order differential equation
>
> $$\ddot{x} + p(t)x = f(t),$$
>
> where $p(t)$ and $f(t)$ are continuous and 2π-periodic, then it will have 2π-periodic solutions if and only if
>
> $$\int_0^{2\pi} \phi(s)f(s)\, ds = 0,$$
>
> where $\phi(t)$ is any 2π-periodic solution of the homogeneous equation.

The proof is not given but the reader is asked to prove a similar result for $p(t) = n^2$, n an integer, in Problem 2. He should also verify that the conditions of the theorem are met in cases 2 and 3 above.

The 2π-periodic solution resulting from the periodic forcing term is called the *forced oscillation* in contrast to the *free oscillation* (not necessarily periodic) which results from the homogeneous equation. It will be useful for later comparison for the reader to consider the equation

$$\ddot{x} + 2b\dot{x} + x = A \cos \omega t, \quad b > 0, \tag{4}$$

and graph the amplitude of the forced oscillation as a function of the frequency ω for various values of b. One obtains something like the diagram shown in Figure 4.2.

For the Equation (4) when $b = 0$ the amplitude is large for ω near 1; resonance occurs when $\omega = 1$. When $b \neq 0$ the gain factor is

$$G = [(1 - \omega^2)^2 + 4b^2\omega^2]^{-1/2},$$

which is smallest when $b = \sqrt{2}/2$—the case of minimum response. For other values of b the maximum amplitude occurs when ω is near 1, the natural frequency, and this fact is useful in the tuning of certain electronic devices.

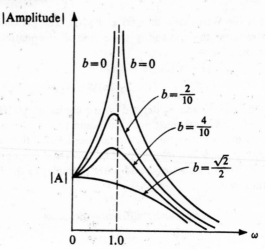

Response Diagram

Figure 4.2

4.3 Conservative Systems

We begin our study of nonlinear equations by examining conservative systems, that is, systems governed by an equation of the form

$$\ddot{x} + f(x) = 0, \tag{5}$$

where we require that

there exists a function $F(x)$ satisfying $F'(x) = f(x)$, hence

$$\int_{x_0}^{x} f(u)\, du = F(x) - F(x_0).$$

These systems occur frequently in applications and have the advantage that a great deal of information about the qualitative behavior of their solutions can often be obtained by graphical rather than analytical techniques.

Equation (5) describes the rectilinear motion of a point with unit mass under the influence of a force depending only on displacement. The velocity of the point is given by $\dot{x}$ and the kinetic energy is $\frac{1}{2}\dot{x}^2$. If we write the equation as $\ddot{x} + F'(x) = 0$, then multiply by $\dot{x}$ and integrate we have

$$\tfrac{1}{2}\dot{x}^2 + F(x) = h, \quad \text{a constant,} \tag{6}$$

where h depends on the initial values at some time t_0 of the displacement and velocity. The relation (6) expresses the law for conservation of energy, namely that kinetic energy + potential energy = constant.

In the xy-plane, the *phase space* of the system governed by (5), the corresponding first order system is

$$\dot{x} = y, \qquad \dot{y} = -f(x). \tag{7}$$

From (6) above we see then that trajectories of (7) satisfy the relation

$$\tfrac{1}{2}y^2 + F(x) = h, \quad \text{a constant,}$$

or

$$y = \pm\sqrt{2}\sqrt{h - F(x)}. \tag{8}$$

Therefore trajectories of (7) are *curves of constant energy*, are symmetric with respect to the x-axis, and have slope at any point given by

$$\frac{dy}{dx} = -\frac{f(x)}{y}.$$

For each value $h = h_0$ for which $h_0 - F(x) \geq 0$ we obtain a trajectory C_0 of the system (7) and it will be defined for all x for which $h_0 - F(x)$ is nonnegative.

A point in the xy-plane for which $\dot{x} = 0$, $\dot{y} = 0$ is called an *equilibrium point* or *critical point* of the system (7). For a conservative system this would represent a point where the velocity and acceleration are zero. We see from (7) that all its equilibrium points are of the form $(x_0, 0)$ where $f(x_0) = 0$, or equivalently $F'(x_0) = 0$. Hence all equilibrium points of (7) lie on the x-axis and correspond to points on the graph of $F(x)$ where it has a local minimum or maximum, or a horizontal point of inflection.

To graph the trajectories of (7) we graph in the xz-plane the graph of the function $z = F(x)$ and the straight line $z = h_0$. As x varies we can compute the difference in the ordinates $h_0 - F(x)$ and use the relation (8) to determine the corresponding values of y. We consider the various cases.

CASE 1: $F(x)$ is monotonic.

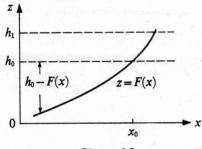

Figure 4.3

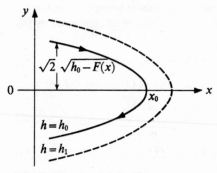

Figure 4.4

As x approaches x_0 for which $F(x_0) = h_0$, the distance $h_0 - F(x)$ and hence y decreases to zero. When $x = x_0$ then $y = 0$ and the trajectory crosses the x-axis at a right angle. Now by symmetry the same diagram is repeated in the lower half plane. The arrow indicates the direction of increasing time and as h varies we get a family of parabola-like trajectories which open to the left if $F(x)$ is increasing and to the right if $F(x)$ is decreasing.

CASE 2: $F(x)$ has a *local minimum*.

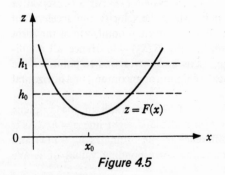

Figure 4.5

From the analysis in the previous case it follows that we will get a family of ellipse-like trajectories in the vicinity of the point $(x_0, 0)$ whenever $F(x)$ has a local minimum at $x = x_0$. Since $F'(x_0) = 0$ the point $(x_0, 0)$ is an equilibrium point of (7) and is called a *center*. We can conclude

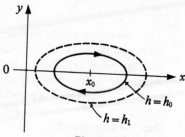

Figure 4.6

that the Equation (5) will have periodic solutions whenever $F(x)$ has a local minimum, i.e., minimum potential energy.

CASE 3: *$F(x)$ has a local maximum.*

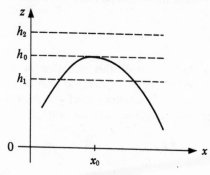

Figure 4.7

In this case if $h_0 = F(x_0)$, the local maximum, then for h near h_0 and $h < h_0$ we will have pairs of parabola-like trajectories crossing the x-axis on opposite sides of x_0 and in opposite directions. For $h > h_0$ there will be pairs of trajectories not crossing the x-axis near x_0. Finally for $h = h_0$ the trajectories will be two curves crossing the x-axis at $x = x_0$ and in opposite directions. Each of the last two trajectories is called a *directrix* and represents solutions $(x(t), y(t))$ of (7) which approach the point $(x_0, 0)$ as t becomes infinite. Since $F'(x_0) = 0$ the point $(x_0, 0)$ is an equilibrium point and is called a *saddle point*.

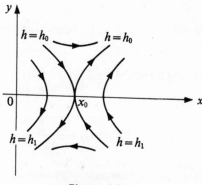

Figure 4.8

OTHER CASES:

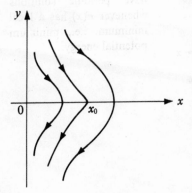

The case where x_0 is a horizontal point of inflection of $F(x)$ is shown and the point $(x_0, 0)$ is called a *degenerate saddle point*. If there exists a value of h for which $h - F(x) > 0$ for $-\infty < x < \infty$ then the corresponding trajectory will be defined for all x and represents a divergent motion. Finally, no trajectories will be defined for any h for which $h - F(x) < 0$.

Figure 4.9

We now examine some nonlinear conservative systems and try to determine regions (if any) where there exist periodic solutions.

Examples

(a) $\ddot{x} + g/l(x - (x^3/6)) = 0$. For $|x|$ small this equation is an approximation of the equation for an undamped pendulum, $\ddot{x} + (g/l) \sin x = 0$, where g is the acceleration of gravity, l is the length and x corresponds to angular displacement. Here

$$f(x) = \frac{g}{l}\left(x - \frac{x^3}{6}\right),$$

so let

$$F(x) = \frac{g}{2l}\left(x^2 - \frac{x^4}{12}\right);$$

proceeding as above we get the diagram shown in Figure 4.10.

Periodic motions will exist for initial conditions

$$|x(t_0)| < \sqrt{6},$$

$$|y(t_0)| = |\dot{x}(t_0)| < \sqrt{\frac{g}{l}}\sqrt{3 - x^2(t_0) + \frac{x^4(t_0)}{12}}.$$

Using the same analysis the reader should compare the above with the results obtained for the actual pendulum equation.

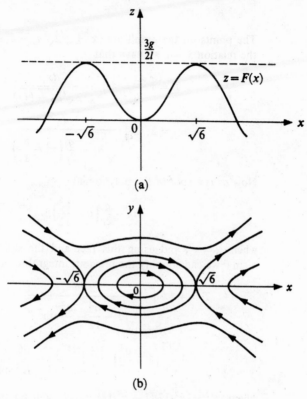

(a)

(b)

Figure 4.10

To find the period T of the periodic solution corresponding to $h = h_0$ and pictured in Figure 4.11, we observe that the period will be four times the time required to go from point A to point B on the graph.

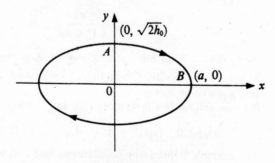

Figure 4.11

The points on the graph are $(x, \pm\sqrt{2}\sqrt{h_0 - F(x)})$, and from the relation $\dot{x} = y$ we have that

$$T = \oint \frac{dx}{y} = 4 \int_0^a \frac{dx}{\sqrt{2}\sqrt{h_0 - F(x)}}$$

$$= 2\sqrt{2} \int_0^a \frac{dx}{\sqrt{h_0 - \frac{g}{2l}\left(x^2 - \frac{x^4}{12}\right)}}.$$

Now a^2 is a root of the polynomial

$$h_0 - \frac{g}{2l}\left(u - \frac{u^2}{12}\right),$$

where $u = x^2$, and using this fact and the substitution $x = a \sin\theta$ we can simplify the previous integral to obtain

$$T = 4\sqrt{\frac{l}{g}} \int_0^{\pi/2} \frac{d\theta}{\sqrt{1 - \frac{a^2}{12} - \frac{a^2}{12}\sin^2\theta}}$$

$$= 4\sqrt{\frac{l}{g}}\left(1 - \frac{a^2}{12}\right)^{-1/2} \int_0^{\pi/2} \frac{d\theta}{\sqrt{1 - K^2\sin^2\theta}},$$

where $K^2 = (a^2/12)(1 - a^2/12)^{-1}$. A first approximation is

$$T \approx 2\pi\sqrt{\frac{l}{g}}\left(1 + \frac{a^2}{24} + \cdots\right),$$

which shows the dependence of the period and hence the frequency of the periodic solution on its amplitude.

(b) $\quad \ddot{x} + \dfrac{x^2(3 - x^4)}{(x^4 + 1)^2} = 0$. The reader can verify that

$$F(x) = \frac{x^3}{x^4 + 1},$$

which has a local minimum at $x = -\sqrt[4]{3}$, a local maximum at $x = +\sqrt[4]{3}$, a point of inflection at $x = 0$, and is asymptotic to the x-axis as $x \to \pm\infty$.
Periodic solutions will then exist for initial conditions

$$x(t_0) < 0, \quad |y(t_0)| = |\dot{x}(t_0)| < \sqrt{2}\sqrt{-F(x(t_0))}.$$

The example is primarily an academic one and was intended to show a possible configuration.

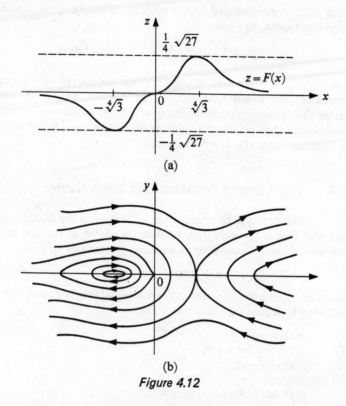

$z = F(x)$

$\frac{1}{4}\sqrt{27}$

$-\sqrt[4]{3}$

$\sqrt[4]{3}$

$-\frac{1}{4}\sqrt{27}$

(a)

(b)

Figure 4.12

Finally a word about stability of equilibrium points. An equilibrium point P is said to be *stable* if

> given any $\epsilon > 0$ there is a δ-neighborhood N_δ of P such that any trajectory starting in N_δ remains in the ϵ-neighborhood of P.

Otherwise it is said to be *unstable*. We see then for conservative systems that a center is stable while a saddle point or degenerate saddle point is unstable.

We may think of the potential energy curve $F(x)$ as describing a frictionless wire with a bead sliding on it. When the bead is resting in a "valley" then a slight push will result in a small periodic motion. Whereas if the bead is resting on a "hill" then a slight push results in its sliding off the hill.

If the wire were not frictionless we might think of this as corresponding to an equation of the form

$$\ddot{x} + 2b\dot{x} + f(x) = 0, \quad b > 0,$$

with friction proportional to the velocity. Multiplying by $\dot{x}$ and integrating from t_0 to t gives

$$[\tfrac{1}{2}\dot{x}(t)^2]_{t_0}^t + [F(x)]_{t_0}^t = -\int_{t_0}^t 2b\dot{x}(s)^2 \, ds.$$

The right side is negative, which says that the energy of the system decreases. We would then expect that damped oscillatory motion rather than periodic motion occurs near a center, while the behavior is unchanged in the vicinity of a saddle point. The reader is asked to verify in Problem 7 that this is indeed the case.

4.4   The Lienard Equation and Limit Cycles

The next type of equation we will discuss is the Lienard equation and it is a generalization of those discussed at the end of the previous section. We consider an equation of the form

$$\ddot{x} + f(x)\dot{x} + g(x) = 0, \tag{9}$$

where we assume that $f(x)$ and $g(x)$ are analytic functions and that the following hypotheses are satisfied:

(i) $g(x) = -g(-x)$;
(ii) $xg(x) > 0$ for $x \neq 0$;
(iii) $f(x) = f(-x)$;
(iv) $f(0) < 0$;
(v) $\int_0^x f(u)du = F(x) \to \infty$ as $x \to \infty$; and
(vi) $F(x) = 0$ has a unique positive root $x = a$.

Now under *all* these hypotheses the reader may think that we should be able to conclude that solutions of (9) turn cartwheels every half hour, but let us examine things more closely.

First of all, the hypotheses (i) and (ii) on $g(x)$ are certainly satisfied by any polynomial or power series with positive coefficients and odd powers of x. If we let

$$G(x) = \int_0^x g(u) \, du,$$

then we can infer that $G(x)$ has an absolute minimum at $x = 0$ and is concave upward. Therefore by our results on conservative systems every solution of (9) would be periodic if there were no damping present.

Secondly the hypotheses (iii) → (v) imply that for $|x|$ near zeor the damping coefficient is negative so motion near zero tends to be oscillatory and increasing in amplitude. But for $|x|$ large, the damping coefficient is positive, so motion tends to be oscillatory but decreasing in amplitude.

If we looked in the phase plane of (9) we might imagine a family of spiral trajectories circling the origin and moving away from it for small amplitude but moving towards it for larger amplitude.

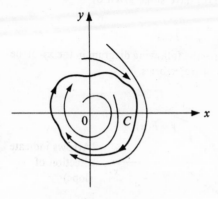

Under these conditions we might suspect that there exists a closed trajectory C separating these two families of trajectories. Such a closed curve would be an example of a *limit cycle*; a closed trajectory representing a periodic solution and having the property that some neighborhood of it contains no other closed trajectory.

In Figure 4.13 above the curve C would represent a *stable* limit cycle inasmuch as solutions near C would tend toward C as t increased. Obviously this is an ideal physical situation since small errors in approximating the periodic solution with a nearby solution would be negligible for large time.

Figure 4.13

Examples of equations of the form (9) and satisfying the hypotheses given are:

(a) $\ddot{x} + \epsilon(x^2 - 1)\dot{x} + x = 0, \quad \epsilon > 0;$
(b) $\ddot{x} - (\dot{x} - \frac{1}{3}\dot{x}^3) + x = 0;$ and
(c) $\ddot{x} + (3x^2 - 2)\dot{x} + x + x^3 = 0.$

Equation (a) is the Van der Pol equation and (b) is Rayleigh's equation which can be put in the form of the Van der Pol equation by letting $\dot{x} = y$ then differentiating. Both these equations arise in the study of certain feedback mechanisms. Equation (c) may be thought of as representing a nonlinear spring with a quadratic damping coefficient.

Finally after all these preliminaries we can proceed to indicate the main points of the proof of the conclusion hinted at above, namely:

> *If the hypotheses (i) through (vi) are satisfied then Equation (9) has a unique stable limit cycle.*

To begin with we can write (9) in the form

$$\frac{d}{dt}(\dot{x} + F(x)) + g(x) = 0,$$

and if we let $y = \dot{x} + F(x)$ then we can consider the equivalent first order system

$$\dot{x} = y - F(x), \qquad \dot{y} = -g(x). \tag{10}$$

The trajectories of the system (10) have slope given by

$$\frac{dy}{dx} = \frac{-g(x)}{y - F(x)},$$

and the hypotheses imply we have the following diagram in the xy-plane for the directions of the slopes of trajectories.

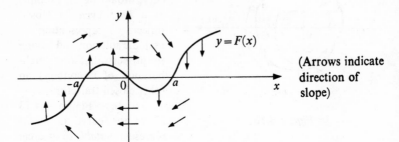

(Arrows indicate direction of slope)

Figure 4.14

Now if a trajectory starts off at $(0, y_1)$ for $y_1 > 0$ then it will proceed in a southeasterly direction and eventually cross the curve $y = F(x)$ since $F(x) \to \infty$ as $x \to \infty$. At this point it will proceed downward and cross the x-axis at say $x = x_0$. To show that it will actually reach the negative y-axis, say at $(0, y_2)$, $y_2 < 0$, note that $g(x)$ and $F(x)$ are bounded on $[0, x_0]$ and $y - F(x)$ becomes large negatively as $y \to -\infty$, hence dy/dx can be made less than -1.

By the same argument and the symmetry of the system (10), a trajectory starting at $(0, -y_1)$ will end up at $(0, -y_2)$, and, since by uniqueness trajectories don't cross each other, it follows that a necessary and sufficient condition for a trajectory to be closed is that $y_1 = -y_2$. The necessity is obvious; as for the sufficiency, observe that if the trajectory going from $(0, y_1)$ to $(0, y_2)$ ends back up at $(0, y_3)$ with $y_3 > 0$; then $y_3 < -y_2 < y_1$. Since the trajectory is closed, then $y_3 = y_1$, hence $-y_2 = y_1$.

Now we introduce the "energy" function

$$L(x, y) = \tfrac{1}{2}y^2 + G(x),$$

where $G(x) = \int_0^x g(u)du$, and clearly by our convention on signs $y_1 = -y_2$ if and only if $L(0, y_2) - L(0, y_1) = 0$. Furthermore

$$L(0, y_2) - L(0, y_1) = \tfrac{1}{2}(y_2^2 - y_1^2)$$
$$= \int_{y_1}^{y_2} y\, dy = \oint_C dL,$$

where the last line integral is evaluated along any curve C from $(0, y_1)$ to $(0, y_2)$ since the first integral is independent of the path chosen.

From the definition of L and Equations (10) we have the following implications:

$$y = \dot{x} + F(x) \Rightarrow dy = \ddot{x}\, dt + f(x)\, dx$$
$$\Rightarrow \dot{x}\, dy = \ddot{x}\, dx + f(x)\dot{x}\, dx,$$

and if (x, y) moves along a trajectory of (10) then

$$dL = L_x\, dx + L_y\, dy = g(x)\, dx + y\, dy$$
$$= g(x)\, dx + (\dot{x} + F(x))\, dy$$
$$= [\ddot{x} + f(x)\dot{x} + g(x)]\, dx + F(x)\, dy$$
$$= 0 + F(x)\, dy.$$

Therefore our problem reduces to finding some trajectory C of (10) for which the line integral

$$\oint_C F(x)\, dy = 0.$$

Each trajectory leaving the positive y-axis crosses the curve $y = F(x)$ at some uniquely determined point $(\alpha, F(\alpha))$ and we can associate α with the trajectory passing through this point. Following the diagram below let

$$\phi(\alpha) = \oint_{ABC} dL = \oint_{ABC} F(x)\, dy.$$

Then our problem is to find α_0 such that $\phi(\alpha_0) = 0$.

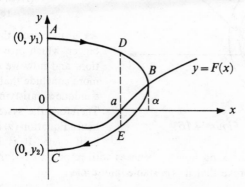

Figure 4.15

By continuous dependence on parameters, ϕ is a continuous function and since $dy < 0$ in the right half-plane and $F(x) < 0$ for $0 < x < a$ we may conclude that α_0, if it exists, satisfies $\alpha_0 \geq a$. We need only consider $\alpha \geq a$.

Now let $\phi(\alpha) = \phi_1(\alpha) + \phi_2(\alpha)$, where

$$\phi_1(\alpha) = \oint_{AD} dL + \oint_{EC} dL, \qquad \phi_2(\alpha) = \oint_{DBE} dL,$$

$$dL = F(x)\,dy.$$

First of all note that $\phi_1(a) > 0$ and we assert that $\phi_1(\alpha)$ decreases as α increases. To see this, consider

$$\oint_{AD} dL = \oint_{AD} f(x)\,dy$$

$$= \oint_{AD} F(x)\,\frac{g(x)}{F(x) - y}\,dx.$$

Then α increasing implies that $|F(x) - y|$ increases, which implies that

$$\oint_{AD} dL \quad \text{as well as} \quad \oint_{EC} dL$$

decreases.

Secondly, note that $\phi_2(a) = 0$ and that $\phi_2(\alpha) < 0$ for $\alpha > a$. We assert that $\phi_2(\alpha)$ decreases as α increases, and to show this make the transformation $X = F(x)$, $Y = y$. Then

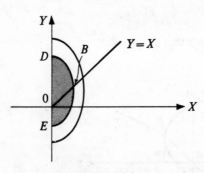

Figure 4.16

$$\phi_2(\alpha) = \int_{DBE} X\,dY$$

$$= -(\text{Area of shaded region}).$$

As α increases the area gets bigger, which proves the assertion, and now we can furthermore conclude that there exists a unique α_0 satisfying $\phi(\alpha_0) = 0$. Therefore the system (10) and hence Equation (9) has a unique limit cycle. It is left for the reader to check on how y_2 behaves with respect to y_1 for $\alpha > \alpha_0$ and $\alpha < \alpha_0$ to prove that it is a stable limit cycle.

Example

For equations of the form

$$\ddot{x} + \epsilon f(x)\dot{x} + x = 0, \quad 0 < \epsilon << 1,$$

where $f(x)$ satisfies the hypotheses above, we would assume that for ϵ very small the limit cycle would be approximated in the phase plane by a circle

$$x = R \cos \theta, \qquad y = R \sin \theta.$$

Along the limit cycle we have

$$\oint_{ABC} F(x) \, dy \simeq \epsilon \int_{-\pi/2}^{\pi/2} F(R \cos \theta) R \cos \theta \, d\theta = 0,$$

which gives us a way of approximating R.

For the Van der Pol equation $f(x) = (x^2 - 1)$ hence

$$F(x) = \left(\frac{x^3}{3} - x\right),$$

and the above relation gives

$$\epsilon \frac{\pi}{2} R^2 \left(\frac{R^2}{4} - 1\right) = 0$$

so $R = 2$. To compute the period we have, since $x^2 + y^2 \cong 4$ and $\dot{y} = -x$, that

$$T = -2 \oint_{ABC} \frac{dy}{x} \cong -2 \int_{-2}^{2} \frac{dy}{\sqrt{4 - y^2}} = 2\pi.$$

So a first approximation to the limit cycle of the Van der Pol equation is $x(t) = 2 \cos(t + \phi)$, ϕ arbitrary, and this is valid for $0 < \epsilon < 0.1$.

To conclude this section it would be interesting to discuss qualitatively what is the nature of the limit cycle of the equation in the previous example when ϵ is large, say $\epsilon \geq 10$. To do so we make a "coordinate stretching" transformation $u = \epsilon y$. Then the equations (10) with $g(x) = x$ become

$$\dot{x} = \epsilon(u - F(x)), \qquad \dot{u} = -\epsilon^{-1} x,$$

or

$$(u - F(x)) \, du = -\epsilon^{-2} x \, dx.$$

Since ϵ^{-2} is very small the trajectories will be approximated by the solutions of $(u - F(x)) \, du = 0$, which are segments of the curve $u = F(x)$ and straight line segments $u = \text{constant}$.

Since we assume that $f(x)$ and $F(x)$ satisfy the hypotheses (iii) → (vi) above, the following approximate picture of a limit cycle will occur. (Figure 4.17). The periodic trajectory will leave, say, the positive y-axis in a horizontal direction until it intersects the curve $u = F(x)$ at B then it will proceed down the curve to the point C. A study of the direction field at point C will convince the reader that the trajectory will now resume its rectilinear motion until it reaches the other branch of the curve $u = F(x)$, etc.

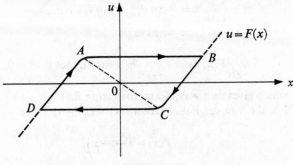

Figure 4.17

Along the lines AB and DC we have that $\dot{u} = 0$, whereas it is decreasing along BC and increasing along DA. If a graph is made of $x(t) = -\epsilon \dot{u}(t)$, one obtains the following picture of a very jerky oscillatory motion (the approximate period of the motion is left for the reader to compute in Problem 11).

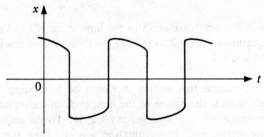

Figure 4.18

This is an example of a *relaxation oscillation* and while the above discussion was heuristic, the existence of such oscillations has been analytically justified by several investigators. These types of oscillation occur in electronic feedback devices as well as in the theory of mechanical shocks.

4.5   The Approximation of Periodic Solutions

To conclude this chapter we wish to discuss the more practical question: Given a nonlinear differential which we know or suspect has a periodic solution, how do we approximate the solution? The methods devised to answer this question are numerous; we have chosen three of them which are sufficiently varied in their approach and also give some insight into other questions such as the stability of the periodic solution.

One type of equation we will consider is the autonomous equation of the form

$$\ddot{x} + x = \epsilon f(x, \dot{x}), \quad 0 < \epsilon < \epsilon_0, \tag{11}$$

where ϵ is a small parameter. When $\epsilon = 0$ all solutions are 2π-periodic so we would suspect that when ϵ is small then periodic solutions of (11) will lie in a neighborhood of some or all of the solutions of the linear equation. This remark is the starting point of many of the methods of approximating periodic solutions of equations of the form (11).

If we add a periodic forcing term $F \cos \omega t$ to the right-hand side of (11), then if $\omega \neq 1$ we know that $2\pi/\omega$-periodic solutions of (11) will exist when $\epsilon = 0$. Again, when ϵ is small, our previous remark applies to this case as well.

Equations of the form (11) are said to be *weakly nonlinear* inasmuch as they represent a small perturbation of a linear equation. Even when an equation is not of the form above it is sometimes useful to introduce a parameter. For example, in the equation representing an approximate pendulum (Example (a) of Section 4.3), if we use initial conditions $x(0) = a$, $\dot{x}(0) = 0$ (the last condition merely shifts the phase of a periodic solution), then near the equilibrium point the periodic solutions can be parametrized in terms of the values of a. Now let $x = ay$ and

$$\tau = \sqrt{\frac{g}{l}}\, t.$$

Then the equation becomes $\ddot{y} + y = \epsilon y^3$ where $\epsilon = a^2/6$ and the differentiation is with respect to τ. With the equation in this form the bookkeeping when computing is much easier.

The Perturbation Method

The basic idea of this method is to assume that the periodic solution(s) of an equation like (11) can be expressed in a power series in the small parameter ϵ, where the coefficients are functions of t. One substitutes the formal power series into the equation and by comparing powers of ϵ determines the coefficients in such a manner that when $\epsilon = 0$ a solution of the linear equation is obtained.

Now if one proceeds as described above the first problem he will encounter is that of *secular terms*—coefficients in the expansion whose amplitude becomes unbounded as t becomes large. A simple example of this phenomenon is the following:

Consider the equation

$$\ddot{x} + (1 + \epsilon)^2 x = 0 \qquad \text{or} \qquad \ddot{x} + x = -\epsilon(2 + \epsilon)x$$

and we will assume initial conditions $x(0) = A$, $\dot{x}(0) = 0$. We suppose that a periodic solution is of the form

$$x(t) = x_0(t) + \epsilon x_1(t) + \epsilon^2 x_2(t) + \cdots,$$

where $x_0(0) = A$, $\dot{x}_0(0) = 0$ and $x_i(0) = \dot{x}_i(0) = 0$ for $i \geq 1$. Substituting the expression above into the differential equation and equating powers of ϵ^n results in the following equations:

$$n = 0: \ddot{x}_0 + x_0 = 0, \quad x_0(0) = A, \quad \dot{x}_0(0) = 0;$$

$$n = 1: \ddot{x}_1 + x_1 = -2x_0, \quad x_1(0) = \dot{x}_1(0) = 0.$$

The solution of the first is $x_0(t) = A \cos t$ and of the second is $x_1(t) = -At \sin t$, so to first order terms

$$x(t) = A \cos t - \epsilon A t \sin t + \cdots.$$

The second term is an example of a secular term—we obtain a poor approximation for large t of the required solution $x(t) = A \cos(1 + \epsilon)t$.

In general the reason for the appearance of secular terms is that the frequency ($\omega = 1$) when $\epsilon = 0$ is not the same ($\omega = 1 + \epsilon$) when $\epsilon \neq 0$. Therefore in using the perturbation one assumes that the frequency of the periodic solution is also expressible as a power series in ϵ with constant coefficients. By an appropriate choice of these coefficients and the amplitudes the secular terms can be eliminated.

Examples

$$\ddot{x} + x = \epsilon x^3, \quad x(0) = A, \quad \dot{x}(0) = 0, \quad 0 < \epsilon < \epsilon_0. \quad (12)$$

Make the substitution $\tau = \omega t$ where

$$\omega = \omega_0 + \epsilon \omega_1 + \epsilon^2 \omega_2 + \cdots,$$

to obtain the equation

$$\omega^2 \ddot{x} + x = \epsilon x^3, \quad \text{(differentiation w/resp. to τ).} \quad (13)$$

For the first equation we suspect a period $T = 2\pi\omega$ where $\omega = 1 + O(\epsilon)$, and with the above substitution, we /now seek periodic solutions of the second equation with period $T = 2\pi$. This follows since $x(\tau) = x(\tau + 2\pi)$ implies

$$x(\omega t) = x(\omega t + 2\pi) = x\left(\omega\left(t + \frac{2\pi}{\omega}\right)\right).$$

Furthermore we are not assuming for now that A is arbitrary: it may turn out that $A = a + O(\epsilon)$ for a specific value of a (e.g. $a = 2$ in the case of the Van der Pol equation).

Now we assume a solution of (13) of the form

$$x(\tau) = x_0(\tau) + \epsilon x_1(\tau) + \epsilon^2 x_2(\tau) + \cdots,$$

where

$$x_i(\tau + 2\pi) = x_i(\tau), \quad \dot{x}_i(0) = 0, \quad i = 0, 1, 2, \ldots.$$

With the expression given for ω we substitute to obtain

$$(\omega_0^2 + 2\epsilon\omega_0\omega_1 + \cdots)(\ddot{x}_0 + \epsilon\ddot{x}_1 + \cdots) + (x_0 + \epsilon x_1 + \cdots)$$
$$= \epsilon(x_0 + \epsilon x_1 + \cdots)^3.$$

Equating powers of ϵ^n gives for $n = 0$

$$\omega_0^2 \ddot{x}_0 + x_0 = 0, \quad \dot{x}_0(0) = 0, \quad x_0(\tau + 2\pi) = x_0(\tau),$$

and the solution is $x_0(\tau) = A_0 \cos \tau$ with $\omega_0 = 1$ and A_0 arbitrary. For $n = 1$ we obtain

$$\omega_0^2 \ddot{x}_1 + x_1 = -2\omega_0\omega_1\ddot{x}_0 + x_0^3, \quad \dot{x}_1(0) = 0,$$
$$\dot{x}_1(\tau + 2\pi) = x_1(\tau),$$

and substituting for x_0 and ω_0 and using the identity

$$\cos^3 \tau = \tfrac{3}{4} \cos \tau + \tfrac{1}{4} \cos 3\tau,$$

we have

$$\ddot{x}_1 + x_1 = (2\omega_1 + \tfrac{3}{4}A_0^2)A_0 \cos \tau + \tfrac{1}{4}A_0^3 \cos 3\tau.$$

There will be no 2π-periodic solutions of this equation unless the expression in parentheses is zero, since a secular term of the form $B\tau \sin \tau$ will occur. To eliminate this term we set $\omega_1 = -\tfrac{3}{8}A_0^2$ and now solve to obtain

$$x_1(\tau) = A_1 \cos \tau - \frac{A_0^3}{32} \cos 3\tau$$

with A_1 arbitrary. To first order in ϵ our solution of (13) is

$$x(\tau) = A_0 \cos \tau + \epsilon\left(A_1 \cos \tau - \frac{A_0^3}{32} \cos 3\tau\right) + \cdots,$$

and since A_0 and A_1 are arbitrary we can satisfy the condition $x(0) = A$ by letting $A_0 = A$ and $A_1 = A^3/32$. We therefore obtain the approximation

$$x(t) = A \cos \omega t + \epsilon \frac{A^3}{32} (\cos \omega t - \cos 3\omega t) + O(\epsilon^2)$$

for the periodic solutions of (12), where

$$\omega = 1 - \frac{3\epsilon}{8} A^2 + O(\epsilon^2).$$

REMARKS: In some discussions of equations of the type above the additional conditions $x_0(0) = A$, $x_i(0) = 0$, $i \geq 1$, are given for the coefficients of the power series. This will work for equations where there is a parametric family of periodic solutions, but fails in the case where there are isolated periodic solutions and each value $x_i(0)$ can only take on specific values. The approach above, where each value of $x_i(0)$ is left initially unspecified, will work in most cases and is to be preferred.

To discuss an equation like

$$\ddot{x} + x = \epsilon x^3 + \epsilon F \cos \omega t$$

using the perturbation method, one introduces the change of variables $\tau = \omega t$ where $\omega = \omega_0 + \epsilon \omega_1 + \cdots$ and obtains the equation

$$\omega^2 \ddot{x} + x = \epsilon x^3 + \epsilon F \cos \tau.$$

Now one proceeds as above looking for solutions $x(\tau) = x_0(\tau) + \epsilon x_1(\tau) + \cdots$ satisfying $x(0) = A$, $\dot{x}(0) = 0$ and $x(\tau + 2\pi) = x(\tau)$. However the method of iteration, to be discussed later, is a much better way of dealing with this type of equation.

The Method of Averaging or Variation of Parameters

If we consider the second order equation $\ddot{x} + x = \epsilon f(x, \dot{x}, t)$, $0 < \epsilon < \epsilon_0$, then the corresponding first order system is

$$\dot{x} = y, \qquad \dot{y} = -x + \epsilon f(x, y, t). \qquad (14)$$

We will assume that f is 2π-periodic in t. When $\epsilon = 0$ the solutions of (14) are of the form

$$x(t) = a \cos t + b \sin t,$$

$$y(t) = -a \sin t + b \cos t$$

where a and b are constants. The method of averaging due to Van der Pol consists of assuming that solutions of the nonlinear equation can be approximated by assuming that a and b are slowly varying functions of time instead. This is analogous to the approach used in solving non-homogeneous linear equations by the method of variation of parameters when solutions of the homogeneous equation are known.

Letting $a = a(t)$ and $b = b(t)$, from the first equation of (14) we have the relation

$$\dot{x} = -a \sin t + b \cos t + \dot{a} \cos t + \dot{b} \sin t$$
$$= y = -a \sin t + b \cos t,$$

and therefore

$$\dot{a} \cos t + \dot{b} \sin t = 0. \tag{15}$$

Now from the second equation of (14) we have

$$\dot{y} = -a \cos t - b \sin t - \dot{a} \sin t + \dot{b} \cos t$$
$$= -a \cos t - b \sin t + \epsilon f(a \sin t + b \cos t, -a \sin t + b \cos t, t)$$

Hence

$$-\dot{a} \sin t + \dot{b} \cos t = \epsilon f(a \sin t + b \cos t, -a \sin t + b \cos t, t), \tag{16}$$

Solving equations (15) and (16) for $\dot{a}$ and $\dot{b}$, we obtain

$$\dot{a} = -\epsilon f(a \cos t + b \sin t, -a \sin t + b \cos t, t)\sin t$$
$$\dot{b} = \epsilon f(a \cos t + b \sin t, -a \sin t + b \cos t, t)\cos t. \tag{17}$$

The system (17) is of the form $\dot{a} = A(a, b, t), \quad \dot{b} = B(a, b, t)$ and we could try to solve it or use numerical analysis but this would generally prove to be very complicated.

We observe instead that by our hypotheses on f the right-hand sides of the equations are 2π-periodic functions of t, so if they did not vary much over a period and ϵ is small then $\dot{a}$ and $\dot{b}$ are small for $0 \le t \le 2\pi$. It would follow then that $a(t)$ and $b(t)$ are slowly varying functions over a period. The approximation will then consist of replacing the right-hand sides of the equations of (17) by their average value over a period, i.e., the first term of their Fourier series expansion.

We thus obtain the equations

$$\dot{a} = \frac{-\epsilon}{2\pi} \int_0^{2\pi} f(a \cos t + b \sin t, -a \sin t + b \cos t, t)\sin t\, dt,$$

$$\dot{b} = \frac{\epsilon}{2\pi} \int_0^{2\pi} f(a \cos t + b \sin t, -a \sin t + b \cos t, t)\cos t\, dt.$$

$$\tag{18}$$

These are the equations of B. Van der Pol and we note that they are of the form of an autonomous system

$$\dot{a} = \epsilon \mathscr{A}(a, b), \qquad \dot{b} = \epsilon \mathscr{B}(a, b).$$

It follows that any values $a = a_0$, $b = b_0$ for which $\mathscr{A}(a_0, b_0) = \mathscr{B}(a_0, b_0) = 0$ will be a solution and

$$x(t) = a_0 \cos t + b_0 \sin t$$
$$y(t) = -a_0 \sin t + b_0 \cos t,$$

will be a first approximation to a period solution of (14).

Let us now assume that f does not depend on t, so the system (14) is autonomous. If we use the polar coordinate representation

$$x = K \cos(t - \theta), \qquad y = -K \sin(t - \theta),$$

where

$$K = \sqrt{a^2 + b^2} \qquad \text{and} \qquad \theta = \tan^{-1} \frac{b}{a},$$

then Equations (18) become

$$\dot{K} = -\frac{\epsilon}{2\pi} \int_0^{2\pi} f(K \cos \mu, -K \sin \mu) \sin \mu \, d\mu = \epsilon \Phi(K),$$

$$\dot{\theta} = \frac{\epsilon}{2\pi K} \int_0^{2\pi} f(K \cos \mu, -K \sin \mu) \cos \mu \, d\mu = \epsilon \Psi(K), \tag{19}$$

where $\mu = t - \theta$. These are the averaging equations due to the mathematicians N. Krylov and N. Bogolyubov and represent a considerable simplification, since the first equation is a first order autonomous equation in K—if we can solve it we can immediately solve for θ.

Suppose that $\Phi(K_0) = 0$; then $K = K_0$ is a solution of the first equation of (19) and the second equation becomes $\dot{\theta} = \epsilon \Psi(K_0)$. If $\Psi(K_0) = 0$, then a first approximation to a periodic solution of $\ddot{x} + x = \epsilon f(x, \dot{x})$ is

$$x(t) = K_0 \cos(t - \theta_0), \quad \theta_0 \text{ constant},$$

and the frequency will be $\omega = 1 + O(\epsilon^2)$. If $\Psi(K_0) \neq 0$, then the approximate periodic solution will be

$$x(t) = K_0 \cos[(1 - \epsilon \Psi(K_0))t - \theta_0], \quad \theta_0 \text{ constant},$$

and the frequency is $\omega = 1 - \epsilon \Psi(K_0) + O(\epsilon^2)$.

If $\Phi(K_0) = 0$ and $K = K_0$ is an isolated zero of Φ, $K_0 \neq 0$, therefore corresponding to a limit cycle, we can examine its stability by letting $K = K_0 + \Delta K$, corresponding to nearby trajectories (if ΔK is small). Then by Taylor's theorem

$$\dot{K} = \dot{\Delta K} = \epsilon \Phi(K_0 + \Delta K)$$
$$= \epsilon \Phi'(K_0) \Delta K + O(\Delta K^2),$$

and if we neglect the higher order terms then

(i) ΔK decreases exponentially if $\Phi'(K_0) < 0$ and the limit cycle is stable;

(ii) ΔK increases exponentially if $\Phi'(K_0) > 0$ and the limit cycle is unstable.

The justification for this conclusion based on only the first order terms can be proved analytically.

Examples

The equation

$$\ddot{x} + x = \epsilon(\alpha + 2\beta x - 3\gamma x^2)\dot{x}, \quad \epsilon, \beta, \gamma > 0$$

describes for instance, a vacuum tube with a tuned grid circuit and a cubic characteristic where x is the ratio of the grid and saturation voltages. The corresponding first order system is

$$\dot{x} = y, \qquad \dot{y} = -x + \epsilon(\alpha + 2\beta x - 3\gamma x^2)y,$$

and using Equations (19) we have

$$\dot{K} = \frac{\epsilon}{2\pi} \int_0^{2\pi} (\alpha + 2\beta K \cos \mu - 3\gamma K^2 \cos^2 \mu) K \sin^2 \mu \, d\mu,$$

$$\dot{\theta} = \frac{-\epsilon}{2\pi K} \int_0^{2\pi} (\alpha + 2\beta K \cos \mu - 3\gamma K^2 \cos^2 \mu) K \sin \mu \cos \mu \, d\mu.$$

Solving this we obtain

$$\dot{K} = \frac{\epsilon K}{2}\left(\alpha - \frac{3\gamma}{4} K^2\right) = \epsilon \Phi(K), \quad \dot{\theta} = 0.$$

We note that $\Phi(0) = 0$ corresponding to an equilibrium point at the origin, and for $\alpha \leq 0$ this will be the only zero of $\Phi(K)$. For $\alpha > 0$ there will be another zero at

$$K_0 = \frac{2}{3}\sqrt{\frac{3\alpha}{\gamma}}$$

and $\Phi'(K_0) = -\alpha < 0$ so this will correspond to a stable limit cycle which can be approximated by

$$x(t) = \frac{2}{3}\sqrt{\frac{3\alpha}{\gamma}} \cos(t - \theta_0).$$

Also $\Phi'(0) = \alpha/2$ so the stability of the origin depends on the value of α—near the origin the equation acts like $\ddot{x} - \epsilon\alpha\dot{x} + x = 0$, describing a damped oscillator.

REMARKS: The above equation is an example of a "soft" excitation inasmuch as small disturbances near the origin when $\alpha > 0$ result in oscillatory motions tending towards a periodic solution. An example of a system with "hard" excitation would be one where the origin is stable and, proceeding from the origin, there is an unstable then a stable limit cycle. One then needs a large initial disturbance from rest to create self-sustaining oscillations tending towards a periodic solution.

Furthermore, note that in the nonautonomous case where $f = f(x, \dot{x}, t)$, the Van der Pol equations (18) are still applicable when f is $2\pi/\omega$-periodic with $\omega \neq 1$. It will only be required that the right-hand sides of equations (17) vary little over $0 \leq t \leq 2\pi$, which will certainly be the case if f is reasonably smooth and ϵ is small. In the nonautonomous case one can also use the polar coordinate representation and equations (19) will be of the form

$$\dot{K} = \epsilon \Phi(K, \theta), \qquad \dot{\theta} = \epsilon \Psi(K, \theta).$$

The Iteration Method

This method is best described by an example, so we will discuss a form of Duffing's equation, namely

$$\ddot{x} + x + \beta x^3 = F \cos \omega t, \quad F > 0. \tag{20}$$

This might describe a spring system with a cubic restoring force and a harmonic driving term. The more general case with a damping term $2b\dot{x}$ will be considered later.

We assume that β is small and rewrite the equation in the form

$$\ddot{x} + \omega^2 x = (\omega^2 - 1)x - \beta x^3 + \beta F_0 \cos \omega t, \tag{21}$$

where $F = \beta F_0$; the reason for this will become clearer in the discussion. For $\beta = 0$ and ω^2 close to 1 an approximate solution will be of the form $x_0(t) = A \cos \omega t$ where A is prescribed. The term $B \sin \omega t$ is omitted since it will disappear at the next step.

Substitute $x_0(t)$ in the right-hand side of (21), reduce the term $\cos^3 \omega t$ and combine terms to obtain

$$\ddot{x} + \omega^2 x = [(\omega^2 - 1)A - \tfrac{3}{4}\beta A^3 + \beta F_0] \cos \omega t - \tfrac{1}{4}\beta A^3 \cos 3\omega t.$$

Since we are looking for periodic solutions we must get rid of the first term on the right-hand side, since it will give rise to a secular term. By setting the term in brackets equal to zero we get the relation

$$\omega^2 = 1 + \tfrac{3}{4}\beta A^2 - \frac{\beta F_0}{A} \quad (\beta F_0 = F), \tag{22}$$

and now we can proceed to solve for $x = x_1(t)$ to obtain

$$x_1(t) = A_1 \cos \omega t + \frac{\beta A^3}{32\omega^2} \cos 3\omega t,$$

where A_1 is arbitrary.

The key feature of this scheme is the relation (22) which may appear strange at first since ω is assumed to be given and A is arbitrary, whereas we have said just the opposite. Actually Equation (22) gives a set of response curves (parametrized by $\beta F_0 = F$) relating ω and A and analogous to those given at the end of Section 4.2 for the linear equation ($\beta = 0$). The device of letting ω depend on A is merely to make the iteration scheme more plausible.

If we wish to continue, the next step in the iteration process is to let $A_1 = A$ in the expression for $x_1(t)$, substitute this in the right-hand side of Equation (21), reduce the powers and products of cosine and proceed as above to get rid of secular terms. This will give an improved version of the relation (22). One should be sure to have an ample supply of patience and scratch paper on hand as well as a good table of trigonometric identities.

To graph the response curves, first graph the parabola $\omega^2 = 1 + \frac{3}{4}\beta A^2$ in the ω^2, A-plane, then graph the family of hyperbolas $\omega^2 = -F/A$ for various values of positive F. Note that if $\beta > 0$ (the case of a "hard" spring force), the parabola opens to the right while if $\beta < 0$ (the case of a "soft" spring force), it opens to the left. Now add the abscissa and finally plot $|A|$ against ω—the sequence of steps is shown below for the case of $\beta > 0$ and several values of F.

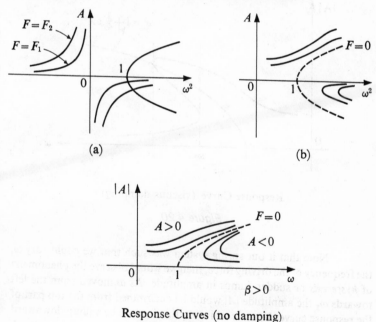

Response Curves (no damping)

(c)

Figure 4.19

For the case $\beta < 0$ there would be a similar picture but bent over to the left.

If we graph the response curves for the linear case ($\beta = 0$) for various values of F we obtain a family of curves rising up from the ω-axis and asymptotic to the vertical line $\omega = 1$ as $|A| \to \infty$. Therefore the effect of the nonlinearity βx^3 is to bend these curves to the right when $\beta > 0$ and to the left when $\beta < 0$. Note that for a given value of ω there may be as many as three values of $|A|$ possible.

We will briefly discuss the situation where viscous damping is present, in which case the Duffing equation is

$$\ddot{x} + 2b\dot{x} + x + \beta x^3 = F \cos \omega t, \quad b, F > 0.$$

For the linear case one obtains families of curves rising up from the ω-axis and peaking near the vertical line $\omega = 1$ (see the diagram at the end of Section 4.2). It is not unreasonable to assume that the effect of the nonlinearity βx^3 will be to bend these curves to the right or left depending on the sign of β. This can be proved to be the case and an example of the type of response curve obtained is shown below for the case $\beta > 0$. For the case $\beta < 0$ the curve is bent to the left.

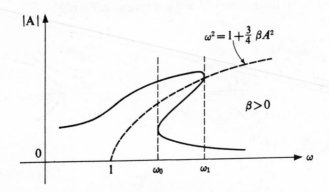

Response Curve (viscous damping)

Figure 4.20

Note that if our spring system was such that we could vary ω, the frequency of the driving force, then we would observe the phenomena of *hysteresis* or sudden jumps in amplitude. As ω moved from the left towards ω_1 the amplitude $|A|$ would be determined from the top part of the response curve. Upon reaching ω_1 there would be a jump downward in amplitude to that given by the lower part of the response curve. A jump upward in amplitude would occur at ω_0 if the frequency were moving from the right past ω_1 towards ω_0. Therefore for frequencies between ω_0 and ω_1 the value of the amplitude depends on the past history of the system. The middle part of the response curve between the two dotted lines is an unstable region.

REMARKS: The notion of a "hard" or "soft" spring comes from the fact that, for a nonlinear spring described by $\ddot{x} + f(x) = 0$, a measure of the stiffness of the spring is given by $f'(x)$. If the stiffness increases with the displacement the spring is said to be *hard*. If it decreases with the displacement, then the spring is said to be *soft*.

To conclude this section we wish to show by an example that it is sometimes advantageous to use more than one method to obtain information about the behavior of solutions of a given equation. The equation we wish to consider is the Mathieu equation.

$$\ddot{x} + (\delta + \epsilon \cos t)x = 0, \quad \delta \geq 0, \tag{23}$$

which along with the Lienard equation is one of the most widely discussed equations in the mathematical literature. We will need the following theoretical result to motivate our investigation:

> *In the δ, ϵ-plane every line $\epsilon = \epsilon_0$, $\epsilon_0 \neq 0$, is divided into alternating intervals of stability and instability of solutions of (23) by a sequence of points (δ_n, ϵ_0), $n = 0, 1, 2, \ldots$, and $\delta_n \to \infty$ as $n \to \infty$. The $\frac{1}{2}$ line of points (δ, ϵ_0), $\delta < \delta_0$ is an interval of instability. Corresponding to each point (δ_n, ϵ_0) there is at least one periodic solution of (23) with period 2π or 4π.*

As we continuously vary the values of ϵ_0 the points $\{\delta_n\}$ will vary continuously so we can imagine the entire δ, ϵ-plane divided into alternate zones of instability by curves $\delta = \delta_i(\epsilon)$, $i = 0, 1, 2, \ldots$, corresponding to values for which 2π or 4π-periodic solutions of (23) exist.*

For $\epsilon = 0$ the only points δ for which 2π or 4π-periodic solutions can exist are the points $\delta = n^2/4$, $n = 0, 1, 2, \ldots$, and these will be boundary points between zones of instability as well as stability. The $\frac{1}{2}$ line $\delta < 0$, $\epsilon = 0$ will consist only of points of instability. We wish to examine the nature of the solutions of (23) in the vicinity of $\delta = \frac{1}{4}$ for small $|\epsilon|$, and to do so we will first use the method of averaging.

Letting $\delta = \frac{1}{4}$ we write the equation in the form

$$\dot{x} = y, \quad \dot{y} = -\tfrac{1}{4}x - \epsilon x \cos t = -\tfrac{1}{4}x + \epsilon f(x, t), \tag{i}$$

---

*   By stability and instability in this case is meant exponential decay or growth of solutions.

and using polar coordinates the approximate solutions will be in the form

$$x(t) = K \cos(\tfrac{1}{2}t - \theta), \quad y(t) = -\tfrac{1}{2}K \sin(\tfrac{1}{2}t - \theta).$$

Prior to averaging one obtains the equations

$$\dot{K} = 2\epsilon[K \cos(\tfrac{1}{2}t - \theta)\cos t]\sin(\tfrac{1}{2}t - \theta)$$
$$= \epsilon K \cos t \sin(t - 2\theta),$$

$$\dot{\theta} = \frac{-2\epsilon}{K}[K \cos(\tfrac{1}{2}t - \theta)\cos t]\cos(\tfrac{1}{2}t - \theta) \qquad (24)$$

$$= -\epsilon[\cos t + \tfrac{1}{2}\cos 2(t - \theta) + \tfrac{1}{2}\cos 2\theta],$$

and averaging the second equation gives $\dot{\theta} = -\tfrac{1}{2}\epsilon \cos 2\theta = \epsilon\Omega(\theta)$, since θ is a constant with respect to integration on t.

Now if $\epsilon > 0$ then the smallest stable zero of $\Omega(\theta)$ is $\theta = -\pi/4$, whereas if $\epsilon < 0$ it is $\theta = \pi/4$. Substituting these values into the first equation of (24) one finds that $\dot{K} = |\epsilon|K \cos^2 t$ and averaging gives the relation $\dot{K} = \tfrac{1}{2}|\epsilon|K$. Therefore an approximate solution for $\delta = \tfrac{1}{4}$ and $|\epsilon|$ small is

$$x(t) = K_0\, e^{|\epsilon|t} \cos\left(\frac{1}{2}t \pm \frac{\pi}{4}\right) \quad (\epsilon \gtrless 0),$$

which grows exponentially as $t \to \infty$ and hence is unstable.

Now to examine the nature of periodic solutions near the points $\delta = 0$ and $\delta = \tfrac{1}{4}$ we will use a perturbation method. Specifically we wish to find approximations to the curves $\delta = \delta(\epsilon)$ associated with the periodic solutions. Therefore let

$$x(t) = x_0(t) + \epsilon x_1(t) + \epsilon^2 x_2(t) + \cdots,$$

$$\delta = \delta_0 + \epsilon\delta_1 + \epsilon^2\delta_2 + \cdots, \qquad (25)$$

where each $x_i(t)$ will be required to have period 2π or 4π, and as $\epsilon \to 0$ these expressions reduce to the relations

$$x(t) = x_0(t) = 1, \qquad\qquad \delta = \delta_0 = 0;$$

$$x(t) = x_0(t) = \begin{cases} \cos \tfrac{1}{2}t, \\ \sin \tfrac{1}{2}t, \end{cases} \qquad \delta = \delta_0 = \tfrac{1}{4}.$$

Now substituting the expressions (25) into Equation (23) and equation powers of ϵ we get the following:

(i) $\delta_0 = 0: \ddot{x}_1 = -\delta_1 - \cos t$. Since $x_1(t)$ must have period 2π or 4π we have that $\delta_1 = 0$ and $x_1(t) = \cos t + \alpha$, α a constant.

$$\ddot{x}_2 = -\delta_2 - x_1(t)\cos t$$
$$= -\delta_2 - \tfrac{1}{2} - \alpha \cos t - \tfrac{1}{2}\cos 2t.$$

The periodicity condition requires that $\delta_2 = -\frac{1}{2}$, hence

$$\delta = \delta(\epsilon) = -\frac{1}{2}\epsilon^2 + \cdots$$

near $\delta = 0$.

(ii) $\delta_0 = \frac{1}{4}$: If $x_0(t) = \cos\frac{1}{2}t$, then we get that

$$\ddot{x}_1 + \frac{1}{4}x_1 = (-\delta_1 - \cos t)\cos\frac{t}{2}$$

$$= (-\delta_1 - \frac{1}{2})\cos\frac{t}{2} - \frac{1}{2}\cos\frac{3}{2}t.$$

For periodicity we must have that $\delta_1 = -\frac{1}{2}$; for the case $x_0(t) = \sin\frac{1}{2}t$ we obtain that $\delta_1 = \frac{1}{2}$ hence

$$\delta(\epsilon) = \frac{1}{4} \pm \frac{1}{2}\epsilon$$

near $\delta = \frac{1}{4}$.

By our previous analysis we know that solutions will be unstable for $\delta = \frac{1}{4}$, $|\epsilon|$ small, and $\delta < 0$, $|\epsilon| = 0$. Using this information and the theoretical result stated above we can sketch the following stability diagram where (i) the shaded regions represent points (δ, ϵ) giving rise to stable solutions of the Mathieu equation; (ii) the unshaded regions are points (δ, ϵ) corresponding to unstable solutions; and (iii) to each point on the boundary curves there corresponds at least one 2π- or 4π-periodic solution.

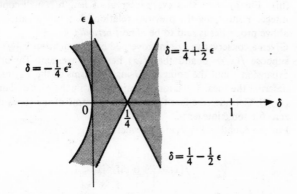

Stability Diagram ($|\epsilon| \ll 1$, $-\infty < \delta < 1$).

Figure 4.21

Problems

1. A system governed by the equation

$$\ddot{x} + 4x = \frac{8}{\pi} \cos 2t, \qquad x(0) = \dot{x}(0) = 0,$$

becomes overloaded (e.g., a fuse blows) when $|x(t)| \geq 12$. Show that this will occur somewhere in the interval $6\pi < t < 25\pi/4$.

2a. Given the equation $\ddot{y} + n^2 y = f(t)$, n an integer and $f(t)$ continuous and 2π-periodic, show that
   (i) periodic solutions will exist if and only if

$$\int_0^{2\pi} f(s)\cos ns \, ds = \int_0^{2\pi} f(s)\sin ns \, ds = 0.$$

   (ii) If the previous condition is not satisfied then all solutions are unbounded.

2b. If $\max|f(t)| = m$, give a precise upper bound for the amplitude of any periodic solution.

3. Given the linear equation $\ddot{x} + 4\pi^2 x = A \sin 2\pi\sqrt{2}\, t$,
   (i) show that there will be a periodic solution only when the initial conditions are $x(0) = 0$, $\dot{x}(0) = -A/4\pi^2$.
   (ii) For any other solution $x(t)$ prove that given any $\epsilon > 0$ there is some integer τ such that

$$|x(t + \tau) - x(\tau)| < \epsilon, \quad -\infty < t < \infty.$$

   Do this by first showing the following: Given any $\epsilon > 0$ there is an integer τ satisfying $|\tau\sqrt{2} - N| < \epsilon/2\pi$ for some integer N.
   (iii) Finally show that every interval of length N will contain an integer τ satisfying the previous relation. A solution satisfying the above properties is said to be *almost periodic*.

4. Given a conservative system governed by the equation $\ddot{x} + f(x) = 0$, suppose $f(x_1) = 0$ and that $f(x)$ has a convergent Taylor series expansion about the point x_1. Using the terms of the Taylor series, describe the local behavior of trajectories in the phase plane near $(x_1, 0)$. Consider the fact that the first n terms of the series may be zero for some integer n.

5. For the equation $\ddot{x} + f(x) = 0$, where

$$f(x) = \begin{cases} -1 & \text{if } x \leq -1 \\ 0 & \text{if } |x| < 1, \\ 1 & \text{if } x \geq 1 \end{cases}$$

sketch the phase plane trajectories, determine the minimum period of oscillation, and find what value of the potential energy yields that period.

6. Given the equation $\ddot{x} + f(x) = 0$,

(i)   Show that for a closed symmetric trajectory about the origin in the phase plane and passing through the point $(a, 0)$, the period is given by

$$T = \sqrt{8} \int_0^a \frac{dx}{\sqrt{\int_x^a f(s)\, ds}}.$$

(ii)   If

$$f(x) = \frac{k}{m} x^{2r+1},$$

r a positive integer, $k, m > 0$, show that all trajectories are periodic and the frequency satisfies the relation

$$\omega^2 = \frac{\pi k}{m} (r+1) a^{2r} \frac{\Gamma^2\left(\dfrac{r+2}{2r+2}\right)}{\Gamma^2\left(\dfrac{1}{2r+2}\right)}$$

where $\Gamma(z)$ is the Gamma function.

7. Given the equation $\ddot{x} + 2b\dot{x} + f(x) = 0$, $b > 0$, where $f(0) = 0$, $f'(0) \neq 0$ and f has a convergent Taylor series near zero. Show that if the equation with no damping has a center at the origin of the phase space then the added damping term will cause trajectories near the origin to spiral in towards it. Show that the behavior in the case of a saddle point is unchanged.

8. For the equation $\ddot{x} + h\dot{x}^2 + x = 0$, $h > 0$, show that
(i)   If $|\dot{x}| = |y|$ is sufficiently small the phase plane will have closed trajectories about the origin for $x < 1/2h$. Find the equation describing the trajectories and show that the curve

$$y^2 = \frac{1}{2h^2} - \frac{x}{h}$$

is a separatrix.
(ii)   If $h > 0$ for $\dot{x} > 0$ and $h < 0$ for $\dot{x} < 0$ (i.e., the friction resists the motion), show that all trajectories spiral towards the origin.

9. Verify that the following equations have at least one limit cycle in the plane (the first two equations are in polar form).

(i) $\dot{r} = r(r^2 - 1)$, $\quad \dot{\theta} = 1$;
(ii) $\dot{r} = r^3 \sin \pi/r$, $\quad \dot{\theta} = 1$;
(iii) $\dot{x} = y$, $\quad \dot{y} = y - x^3 - x^4 y$.

Discuss the stability of the limit cycles.

10. Prove that the unique limit cycle of the Lienard equation is stable.

11. For the Van der Pol equation with large ϵ show that the approximate period of the relaxation oscillation is $T = 1.614\epsilon$.

12. Given the equation $\ddot{x} + x + x^2 = 0$, determine precise estimates on the initial conditions which give rise to a periodic solution. Then use a perturbation expansion to approximate a periodic solution and its frequency up to second order terms.

13.   Use the perturbation method to
(i)   show that an approximation to the limit cycle for the Van der Pol equation with small ϵ is

$$x(t) = 2 \cos \tau + \epsilon(\tfrac{3}{4} \sin \tau - \tfrac{1}{4} \sin 3\tau) + O(\epsilon^2),$$

where $\tau = \omega t$ with

$$\omega = 1 - \tfrac{1}{16}\epsilon^2 + O(\epsilon^3).$$

(ii)   Find the stability curves in the δ, ϵ plane for the Mathieu equation up to and including second order terms near the points $\delta = 0$, $\tfrac{1}{4}$, $1 \tfrac{2}{4}$ and 4 with $|\epsilon|$ small. Draw a sketch of the stability diagram.

14.   Use the method of averaging to find
(i)   an approximate value of the frequency of oscillation for the approximate pendulum equation

$$\ddot{x} + \frac{g}{l}\left(x - \frac{x^3}{6}\right) = 0;$$

(ii)   an approximate solution for large t of the Bessel equation of order zero, $\ddot{x} + t^{-1}\dot{x} + x = 0$. Compare your result with those obtained by a regular asymptotic expansion.

15.   Given the equation

$$\ddot{x} + x = [-\alpha + \beta\dot{x}^2 - \gamma\dot{x}^4]\dot{x}, \quad \beta, \gamma > 0,$$

use the method of averaging in the phase plane to show that
(i)   if $\alpha > 9\beta^2/40\gamma$, then the origin will be a stable equilibrium point and no limit cycles occur;
(ii)   if $9\beta^2/40\gamma > \alpha > 0$, then there will be two limit cycles and hard self excitation occurs.
(iii)   if $\alpha < 0$ the origin is unstable and there is one stable limit cycle—a case of soft self excitation.

16.   An alternate form of iteration for the Duffing equation is to write it in the form

$$\ddot{x} = -x - \beta x^3 + F \cos \omega t,$$

substitute $x = x_0(t) = A \cos \omega t$ in the right-hand side, solve for $x = x_1(t)$, then require that the coefficient of the $\cos \omega t$ term be A. Use this method to find a first approximate solution $x_1(t)$ of

$$\ddot{x} + x + \beta x^3 = F_1 \cos \omega_1 t + F_2 \cos \omega_2 t, \quad F_1, F_2 > 0,$$

where $x_0(t) = A \cos \omega_1 t + B \cos \omega_2 t$. Give the relations between A, B, ω_1 and ω_2 corresponding to Equation (22). Do terms with frequencies other than integer multiples of ω_1 and ω_2 appear?

17.   Given the equation $\ddot{x} + p(t)x = 0$, where

$$p(t) = \begin{cases} \omega^2 + \alpha^2 & \text{if } 0 \le t < \pi \\ \omega^2 - \alpha^2 & \text{if } \pi \le t < 2\pi, \end{cases} \quad p(t + 2\pi) = p(t),$$

construct a stability diagram in the ω^2, α^2-plane for $0 \le \omega^2 \le 4$ and α^2 small. Compare it to that obtained for the Mathieu equation.

18.   Given a nonlinear equation

$$\ddot{x} + f(x, \dot{x}) + g(x) = F(t)$$

and an approximate solution $x = x_a(t)$, one can study its stability by introducing a small perturbation and letting $x = x_a(t) + \eta(t)$;

(i)   assuming that f and g are analytic and that $x_a(t)$ satisfies the equation obtain the *variational equation*

$$\ddot{\eta} + \left(\frac{\partial f}{\partial \dot{x}}\right)_{x=x_a} \dot{\eta} + \left(\frac{\partial f}{\partial x} + \frac{\partial g}{\partial x}\right)_{x=x_a} \eta = 0.$$

One can now study the stability of solutions of the variational equation (which is linear) to determine the stability of $x_a(t)$.

(ii)   Obtain the variational equation for the Duffing equation $\ddot{x} + x + \beta x^3 = \beta F_0 \cos \omega t$ where $x_a(t) = A \cos \omega t$. Show that it can be transformed into a Mathieu equation (23) with

$$\delta = \frac{1}{\omega^2}\left(\frac{1}{4} + \frac{3}{8}\beta A^2\right), \qquad \epsilon = \frac{3}{8}\frac{\beta A^2}{\omega^2}.$$

(iii)   Use the frequency amplitude relation (22) to show that if β is small, then to first order terms in β

$$\delta = \frac{1}{4} + \frac{3}{16}\beta A^2 + \frac{1}{4}\beta \frac{F_0}{A}, \qquad \epsilon = \frac{3}{8}\beta A^2.$$

(iv)   If $\omega > 1$, hence three values of A are possible, use the stability diagram for the Mathieu equation to show that the largest and smallest values of A give stable solutions of Duffing's equation, whereas the intermediate value gives unstable solutions.

References

1.   A. A. Andronow and C. E. Chaikin, *Theory of Oscillations*, Princeton University Press, Princeton, N. J., 1949.

2.   F. M. Arscott, *Periodic Differential Equations*, Pergamon Press, Oxford, 1964.

3.   N. N. Bogolyubov and Y. A. Mitropolsky, *Asymptotic Methods in the Theory of Oscillations*, Gordon & Breach, New York, 1961.

4.   F. Brauer and J. A. Nohel, *Qualitative Theory of Ordinary Differential Equations*, W. A. Benjamin, New York, 1969.

5.   N. V. Butenin, *Elements of the Theory of Nonlinear Oscillations*, Blaisdell, New York, 1965.

6.   L. Cesari, *Asymptotic Behavior and Stability Problems in Ordinary Differential Equations*, 2nd Edition, Academic Press, New York, 1963.

7.   W. J. Cunningham, *Introduction to Nonlinear Analysis*, McGraw-Hill, New York, 1958.

8.   T. V. Davies and E. M. James, *Nonlinear Differential Equations*, Addison-Wesley, Reading, Mass., 1966.

9.   J. K. Hale, *Oscillations in Nonlinear Systems*, McGraw-Hill, New York, 1963.

10.   N. Minorsky, *Introduction to Non-Linear Mechanics*, J. W. Edwards, Ann Arbor, Mich., 1946.

11.   J. J. Stoker, *Nonlinear Vibrations in Mechanical and Electrical Systems*, Interscience, New York, 1950.

12.   R. A. Struble, *Nonlinear Differential Equations*, McGraw-Hill, New York, 1962.

13.   M. Urabe, *Nonlinear Autonomous Oscillations; Analytic Theory.* Academic Press, New York, 1967.

5

Boundary Value Problems and the Calculus of Variations

5.1 Introduction

In the previous chapters we have discussed some types of boundary value problems of the general form: Given the second order differential equation $y'' = f(x, y, y')$, find a solution $y(x)$ satisfying two distinct endpoint conditions. If $a < b$, then examples of such conditions would be

(i) $y(a) = A$, $y(b) = B$;
(ii) $y(a) = A$, $y'(b) = C$;
(iii) $p_1 y'(a) + q_1 y(a) = 0$, $p_2 y'(b) + q_2 y(b) = 0$;

where at least one of the constants p_1, q_1 and one of p_2, q_2 are not zero. Even the search for a periodic solution may be thought of as a special type of boundary value problem where f is 2π-periodic in x and a solution must satisfy $y(0) = y(2\pi)$, $y'(0) = y'(2\pi)$.

Up to now we have dealt with problems of this type by either showing directly the existence of the required solution or by expansion in a series involving the independent variable and/or some parameters. In this chapter we will take another approach, namely that the solutions of certain types of boundary value problems will be regarded as extremals of certain integrals to be minimized which depend on an unknown function. In the process of approximating the minimum we will hope to obtain an approximation to the solution or alternatively to the eigenfunctions and associated eigenvalues of the boundary value problem.

This approach is often referred to as a variational approach or method because of its relation to the calculus of variations, one of the principal areas of study in analysis and applied mathematics. We cannot hope to cover even a modicum of this subject here, but will hopefully be able to indicate to the reader its relationship to boundary value problems.

5.2 The Simplest Problem of the Calculus of Variations

The reader will recall from his calculus days that given a function $y(x)$, $a \leq x \leq b$, with a continuous (or even piecewise continuous*) derivative that its arc length is given by the integral

$$I[y] = \int_a^b \sqrt{1 + y'(x)^2}\, dx.$$

Suppose now that we are given a family K of such functions, then to each member $y = y(x)$ in K the integral $I[y]$ assigns a real number—its arc length. We can then ask the question whether there is an element of K which gives $I[y]$ a minimum in the family? The answer clearly depends on the family K as the following examples show.

Examples

(a) Let K be the set of all functions $y = y(x)$, $-1 \leq x \leq 1$, $-\infty < y < \infty$, having piecewise continuous first derivatives, and satisfying $y(-1) = y(1) = 0$. Then the straight line $y(x) \equiv 0$, $-1 \leq x \leq 1$, gives $I[y]$ a minimum in K, i.e., is that function in K having the smallest arc length.

(b) Now suppose we add the requirement that all the functions in K satisfy the inequality $x^2 + y^2(x) \geq \frac{1}{2}$, i.e. they cannot pass through the interior of the circles $x^2 + y^2 = \frac{1}{2}$. Then there are two functions in K which minimize $I[y]$ as shown in Fig. 5.1.

---

* By piecewise continuous we mean having at most a finite number of jump discontinuities in the interval.

Finally if we require that $x^2 + y^2(x) > 1$, then clearly no member of K gives $I[y]$ a minimum, but certainly $I[y]$ has an (unattainable) greatest lower bound in K.

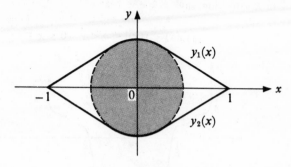

Figure 5.1

The two examples above illustrate the following type of problem: one is given a family K of functions $y = y(x)$, $a \le x \le b$, all passing through two given points (a, y_a), (b, y_b), and an integral of the form

$$I[y] = \int_a^b f(x, y(x), y'(x)) \, dx,$$

defined for the elements of K. One asks the question whether there is a function which gives $I[y]$ an absolute minimum in K?

As in the case of minima of a function of a real variable, one may be only able to find relative minima as the following examples illustrate.

(c)   Let K be the family of all functions $y = y(x)$, $0 \le x \le 1$, satisfying $y(0) = y(1) = 1$ and having piecewise continuous first derivatives. Let

$$I[y] = \int_0^1 y(x)y'(x)^2 \, dx,$$

and it is clear that for $y(x)$ positive on $0 \le x \le 1$ that $I[y]$ is nonnegative. Clearly the function $y_0(x) \equiv 1$, $0 \le x \le 1$ belongs to K and $I[y_0] = 0$. But if $y(x)$ is allowed to become negative then $I[y]$ can be made as large negatively as desired. For instance, for the function $y_\epsilon(x)$ (Figure 5.2) one can show that

$$\int_{1/4}^{3/4} y_\epsilon(x)y_\epsilon'(x)^2 \, dx = -N^3\epsilon^2,$$

so that $I[y_\epsilon]$ can be made as large negatively as desired. Therefore the curve $y_0(x) \equiv 1$, $0 \le x \le 1$, is a relative minimum in K in the sense that $I[y_0] \le I[y]$ for any $y = y(x)$ in K satisfying, say,

$$|y_0(x) - y(x)| < \delta < 1, \quad 0 \le x \le 1.$$

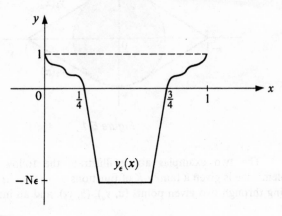

$$0 < \epsilon < \tfrac{1}{4}, \; N \gg 0$$

Figure 5.2

(d) Let K be the family of all functions $y = y(x)$, $1 \le x \le 2$ satisfying $y(1) = 1$, $y(2) = 4$, and having piecewise continuous first derivatives. Let

$$I[y] = \int_1^2 x^3 y'(x)^{-2} \, dx,$$

and it can be shown that $y_0(x) = x^2$ gives a relative minimum (the proof of this is beyond the intent of this book) and $I[y_0] = \tfrac{3}{2}$. Clearly, in this case the value of the derivative $y'(x)$ cannot be too large or $I[y]$ can be made as close to zero as desired. Therefore $I[y_0] \le I[y]$ for any $y = y(x)$ in K satisfying

$$|y_0(x) - y(x)| < \delta_1, \; |y_0'(x) - y'(x)| < \delta_2, \quad 1 \le x \le 2,$$

where δ_1 and δ_2 are certain fixed positive numbers. For instance if one "approximates" $y_0(x) = x^2$ with a sawtoothed curve $y_k(x)$ having slope $\pm k$, then

$$I[y_k] = \int_1^2 \frac{x^3}{k^2} \, dx = \frac{15}{4k^2}$$

and $I[y_k] \geq \frac{3}{8}$ only if $k^2 \leq 10$. Using this fact one can obtain the rough estimate $\delta_2 < 4 + \sqrt{10}$. Note that

$$|y_k(x) - y_0(x)|$$

can be made as small as desired by increasing the number of "teeth" of $y_k(x)$. To summarize: we are given

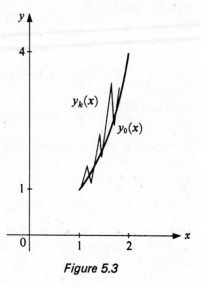

Figure 5.3

(i) a family K of functions $y = y(x)$, $a \leq x \leq b$, satisfying $y(a) = y_a$, $y(b) = y_b$ where y_a and y_b are fixed, and having piecewise continuous first derivatives;

(ii) an integral

$$I[y] = \int_a^b f(x, y(x), y'(x))\, dx$$

defined on the family K. Such an integral is an example of a *functional*, that is, a mapping from a family of functions into the real numbers.

The so-called simplest or classical problem of the calculus of variations is to determine whether there are any elements of K which give the functional $I[y]$ an absolute or relative minimum (or maximum) in the family K. The word "simplest" should not be misconstrued, inasmuch as a list of mathematicians who contributed to this problem would include such luminaries as the Bernoulli brothers, Lagrange, Euler, Weierstrass, Legendre, Jacobi, Tonelli, and Hilbert.

5.3 The Euler-Lagrange Equation

In this section we will derive a necessary condition in the form of an ordinary differential equation which a function in K must satisfy if it is an absolute or relative minimum for $I[y]$. Since all the functions in K satisfy the condition $y(a) = y_a$, $y(b) = y_b$, then that function which gives $I(y)$ a minimum is consequently the solution of a boundary value problem. We will investigate this relation further in the next sections.

First of all to avoid repetition let us refer to the class of functions which are continuous and have piecewise continuous derivatives on $[a, b]$ as the class $C^1_g [a, b]$. Therefore the class K is the set of all functions $y(x)$ in $C^1_g [a, b]$ which satisfy $y(a) = y_a$, $y(b) = y_b$.

Secondly, we will need the following result to establish the necessary condition:

> If $M(x)$, $a \leq x \leq b$, is a given piecewise continuous function and
> $$\int_a^b M(x)\eta(x) \, dx \equiv 0$$
> for every continuous function $\eta(x)$ satisfying $\eta(a) = \eta(b) = 0$, then $M(x) \equiv 0$ in $[a, b]$.

The proof is by contradiction. For if $M(x) \not\equiv 0$, then there must be some interval $[\alpha, \beta]$ in $[a, b]$ in which say $M(x)$ is strictly positive. Now let $\eta_0(x) = (x - \alpha)(\beta - x)$ for $\alpha \leq x \leq \beta$ and zero elsewhere, then

$$\int_a^b M(x)\eta_0(x) \, dx > 0$$

which gives the desired contradiction. Clearly the same proof would apply if we required that all the functions $\eta(x)$ belong to $C^1_g [a, b]$ instead.

Given the class K of functions described above and the functional

$$I[y] = \int_a^b f(x, y(x), y'(x)) \, dx,$$

let us assume that

(i)   the integrand $f(x, y, y')$ regarded as a real valued function of three real variables has continuous first and second partial derivatives, and

(ii)   that there is a function $y_0(x)$ in K which gives $I[y]$ a relative or absolute minimum, and

(iii)   that $y_0(x)$ has a piecewise continuous second derivative.

This last assumption is a strong one but will be justified when we discuss those functionals related to certain boundary value problems.

Let $\eta(x)$ be *any* function in $C^1_g[a, b]$ and satisfying $\eta(a) = \eta(b) = 0$. Then the function $y(x) = y_0(x) + t\eta(x)$ belongs to K for any value of the parameter t. For $|t|$ sufficiently small the quantities $|y_0(x) - y(x)|$ and $|y'_0(x) - y'(x)|$ can be made as small as desired and so by our assumption $I[y] \geq I[y_0]$ for $|t|$ sufficiently small.

Now let us consider the function

$$\phi(t) = \int_a^b f(x, y(x), y'(x)) \, dx$$
$$= \int_a^b f(x, y_0(x) + t\eta(x), y'_0(x) + t\eta'(x)) \, dx.$$

We see that it is defined for t near zero and has a local minimum at $t = 0$. It is also a continuously differentiable function near $t = 0$ and by the chain rule, its derivative is given by the expression

$$\phi'(t) = \int_a^b \left[\frac{\partial f}{\partial y} \eta(x) + \frac{\partial f}{\partial y'} \eta'(x) \right] dx,$$

where the arguments of the partial derivatives are x, $y(x)$, and $y'(x)$.

We can then conclude that $\phi'(0) = 0$, which implies that the previous expression is zero when the arguments of the partial derivatives are x, $y_0(x)$ and $y_0'(x)$, i.e.,

$$\int_a^b \left[\frac{\partial f(x, y_0(x), y_0'(x))}{\partial y} \eta(x) + \frac{\partial f(x, y_0(x), y_0'(x))}{\partial y'} \eta'(x) \right] dx = 0.$$

The second expression in the integrand is integrable by parts and since $\eta(a) = \eta(b) = 0$, we obtain the expression

$$\int_a^b \left[\frac{\partial f(x, y_0(x), y_0'(x))}{\partial y} - \frac{d}{dx} \frac{\partial f(x, y_0(x), y_0'(x))}{\partial y'} \right] \eta(x) \, dx = 0,$$

and this is true for every such $\eta(x)$. We can conclude from this and the previous result that for $a \leq x \leq b$,

$$\frac{\partial f(x, y_0(x), y_0'(x))}{\partial y} - \frac{d}{dx} \frac{\partial f(x, y_0(x), y_0'(x))}{\partial y'} = 0.$$

This follows since by our assumptions the expression in brackets is piecewise continuous (the reader should verify this).

We can summarize all the above as follows:

> If $y = y_0(x)$ gives an absolute or relative minimum to $I[y]$ in the class K then $y_0(x)$ is a solution of the Euler-Lagrange equation
>
> $$\frac{\partial f}{\partial y} - \frac{d}{dx} \frac{\partial f}{\partial y'} = 0.$$

Some important observations should be made:

(i) While the above relation was derived assuming that $y_0(x)$ possessed a piecewise continuous second derivative it also holds at every point of continuity of $y_0'(x)$ when $y_0(x)$ belongs to the class $C_g^1[a, b]$.

(ii) Since $f = f(x, y, y')$, the Euler-Lagrange equation is an ordinary differential equation. The function $y_0(x)$ must satisfy the boundary conditions $y(a) = y_a$, $y(b) = y_b$, so we see that the minimizing function is the solution of a boundary value problem.

(iii)   The Euler-Lagrange equation is a *necessary* condition for a minimum. Given a functional $I(y)$ of the form above, we can certainly obtain the Euler-Lagrange equation and may even be able to find its solution(s). Whether $I[y]$ has a minimum and whether any of the solutions of the Euler-Lagrange equation furnish this minimum, it remains to be proved or justified by physical reasoning (or faith).

Some examples follow and as is the custom, we will omit the dependence on x in the arguments $y(x)$ and $y'(x)$ of the integrand f.

(a)
$$I[y] = \int_a^b \sqrt{1 + y'^2}\, dx,$$

the arc length integral.

Then $f(x, y, y') = \sqrt{1 + y'^2}$ and

$$\frac{\partial f}{\partial y} = 0, \qquad \frac{\partial f}{\partial y'} = \frac{y'}{\sqrt{1 + y'^2}},$$

so the Euler-Lagrange equation is

$$0 - \frac{d}{dx}\left(\frac{y'}{\sqrt{1 + y'^2}}\right) = 0,$$

or

$$\frac{y'}{\sqrt{1 + y'^2}} = C, \quad \text{a constant.}$$

A solution is $y' = $ constant or $y(x) = Ax + B$, with A and B constant. Thus a straight line is a solution and the constants A and B must be determined from the boundary conditions.

(b)
$$I[y] = \int_a^b yy'^2\, dx,$$

(Example (c) of the last section). Here $f(x, y, y') = yy'^2$, so

$$\frac{\partial f}{\partial y} = y'^2, \qquad \frac{\partial f}{\partial y'} = 2yy',$$

and

$$\frac{d}{dx}\frac{\partial f}{\partial y'} = 2y'^2 + 2yy''.$$

The Euler-Lagrange equation is then

$$2yy'' + y'^2 = 0,$$

and one solution is $y' = 0$ or $y(x) = $ constant. If we assume that $y' \neq 0$ and $y > 0$, then the equation is equivalent to the relation $d/dx(2yy'^2) = 0$, or $yy'^2 = C$, a positive constant. The solutions then are of the form $y(x) = B|x - A|^{2/3}$, when A and $B > 0$ are constants to be determined.

(c)
$$I[y] = \int_0^1 xy'^2 \, dx,$$

and K is the family of functions $y(x)$ in $C_g^1[0, 1]$ satisfying $y(0) = 1$ and $y(1) = 0$. The Euler-Lagrange equation is

$$\frac{d}{dx}(2xy') = 0 \quad \text{or} \quad 2xy' = C, \quad \text{a constant.}$$

If $C \neq 0$, one can solve the equation or merely note that it implies that

$$I[y] = \frac{C^2}{4} \int_0^1 \frac{dx}{x},$$

which diverges. If $C = 0$, then the boundary conditions cannot be satisfied—we may suspect that $I[y]$ has no minimum in K, which is indeed the case (see Problem 1).

Again we note that in Examples (a) and (b) the functions are solutions of the Euler-Lagrange equation and it remains to be proved that they give a minimum to the respective functionals $I[y]$.

5.4   The Boundary Value Problem

We now consider the following functional:

$$I[y] = \int_a^b [p(x)y'^2 + q(x)y^2 + 2f(x)y] \, dx, \tag{1}$$

where

(i) $p(x)$ is continuously differentiable on $a \leq x \leq b$,
(ii) $q(x)$ and $f(x)$ are continuous on $a \leq x \leq b$, and
(iii)   the class K is the set of all functions $y(x)$ in $C_g^1[a, b]$ satisfying the boundary conditions $y(a) = y_a$, $y(b) = y_b$, with y_a and y_b fixed. Then

$$f(x, y, y') = p(x)y'^2 + q(x)y^2 + 2f(x)y,$$

and the Euler-Lagrange equation is

$$\frac{d}{dx}(p(x)y') - q(x)y - f(x) = 0. \tag{2}$$

Thus we see that one of the possible candidates in K which gives $I[y]$ a minimum (relative or absolute) is the solution of (2)—the second order linear equation in self-adjoint form—satisfying the boundary condition $y(a) = y_a$, $y(b) = y_b$.

The reader may have noted one slight discrepancy, namely that while we are looking for a minimum of the functional (1) in the class K of functions in $C_g^1[a, b]$, the Euler-Lagrange (2) equation is a second order equation whose solution must have a second derivative—in fact a continuous one. This matter is settled in most cases by the following result:

> If $p(x) > 0$ and $q(x) \geq 0$ for $a \leq x \leq b$,
> than the solution of the Euler-Lagrange equation
> (2) is a unique absolute minimum for the functional
> $I[y]$ in K.

To prove this, suppose $y_0(x)$ is the solution and $y(x)$ is any other function in K. Then we can write $y(x) = y_0(x) + \eta(x)$ and $\eta(x)$ is in $C_g^1[a, b]$. If one computes the difference between $I[y]$ and $I[y_0]$, one obtains, after integration by parts,

$$I[y] - I[y_0]$$

$$= I[y_0 + \eta] - I[y_0]$$

$$= 2 \int_a^b [p(x)y_0'(x)\eta'(x) + q(x)y_0(x)\eta(x) + f(x)\eta(x)]\,dx$$

$$+ \int_a^b [p(x)\eta'(x)^2 + q(x)\eta(x)^2]\,dx$$

$$= 2 \int_a^b \left[-\frac{d}{dx}(p(x)y_0'(x)) + q(x)y_0(x) + f(x) \right]\eta(x)\,dx$$

$$+ \int_a^b [p(x)\eta'(x)^2 + q(x)\eta(x)^2]\,dx.$$

The first expression is zero, since $y_0(x)$ is a solution of (2), and the second expression is positive, hence $I[y_0] < I[y]$ so $y_0(x)$ gives $I[y]$ an absolute minimum. The uniqueness is left to the reader to prove.

Now consider the functional

$$I[y] = \int_a^b [p(x)y'^2 + (q(x) - \lambda r(x))y^2]\,dx, \tag{3}$$

where λ is a real or complex parameter; then the Euler-Lagrange equation is

$$\frac{d}{dx}(p(x)y') - q(x)y + \lambda r(x)y = 0. \tag{4}$$

In this case we have an eigenvalue problem of the Sturm-Liouville type.

Suppose that $y_a = y_b = 0$, and that $p(x)$ is positive and continuously differentiable, $r(x)$ is positive and continuous, and $q(x)$ is continuous on a $a \leq x \leq b$. Then the theory of the Sturm-Liouville problem tells us that there exist

(i) a countable sequence of eigenvalues λ_n, $n = 1, 2, \ldots$, satisfying $\lambda_1 < \lambda_2 < \cdots < \lambda_n \cdots$, and $\lim_{n \to \infty} \lambda_n = \infty$;

(ii) a countable sequence of eigenfunctions $y_n(x)$, $n = 1, 2, \ldots$, such that $y_n(x)$ vanishes exactly n times on $a \leq x \leq b$ and is a solution of (4) when $\lambda = \lambda_n$.

In addition we have the orthonormality relations

$$\int_a^b r(x)y_n^2(x)\,dx = 1, \qquad \int_a^b r(x)y_n(x)y_m(x)\,dx = 0, \quad m \neq n$$

(the first relation is obtained by an appropriate normalization).

Now consider the functional

$$I_1[y] = \int_a^b [p(x)y'^2 + q(x)y^2]\,dx, \tag{5}$$

and it can be shown that it has an absolute minimum in the subclass K_1 of functions of K satisfying $\int_a^b r(x)y^2(x)\,dx = 1$, and that the minimizing function $\tilde{y}(x)$ is continuously twice differentiable. If $I_1[\tilde{y}] = \tilde{\lambda}$, then for any $y = y(x)$ in K_1 we have that

$$I_1[y] \geq I_1[\tilde{y}] = \tilde{\lambda},$$

so in fact $\tilde{y}(x)$ gives an absolute minimum (equal to zero) to the functional

$$I_1[y] - \tilde{\lambda}\int_a^b r(x)y^2(x)\,dx$$

in K_1. But this is just the functional (3) with $\lambda = \tilde{\lambda}$ and since $\tilde{\lambda}$ is the smallest value this implies the following:

> Under the hypotheses given, the first eigenfunction $y_1(x)$ of Equation (4) gives an absolute minimum to the functional (5) in the class of all functions in $C_g^1[a, b]$ satisfying $y(a) = y(b) = 0$ and $\int_a^b r(x)y^2(x)\,dx = 1$. Furthermore $I_1[y] = \lambda_1$, the first eigenvalue of (4).

A similar line of reasoning shows that $y_n(x)$, the nth eigenfunction of (4), gives an absolute minimum to $I_1[y]$ in the class of functions in

$C_g^1[a, b]$ satisfying $y(a) = y(b) = 0$, $\int_a^b r(x)y^2(x)\, dx = 1$, and $\int_a^b r(x)y_j(x)y(x)\, dx = 0$, $j = 1, 2, \ldots, n-1$. Furthermore $I_1[y_n] = \lambda_n$, the nth eigenvalue of (4).

Examples

(a)
$$I[y] = \int_a^b [y'^2 - y^2]\, dx.$$

Then $p(x) \equiv 1$, $q(x) = -1$ and $f(x) \equiv 0$ so the Euler-Lagrange equation is the familiar $y'' + y = 0$. Its solutions are $y(x) = A \cos x + B \sin x$ and if $b - a < \pi$ it is easy to show one can uniquely solve for A and B so that $y(a) = y_a$, $y(b) = y_b$. Therefore the Euler-Lagrange equation furnishes only one possible candidate to minimize $I[y]$ in K.

(b)
$$I[y] = \int_1^2 \left[xy'^2 + \frac{1-x^2}{x}\, y^2 - 2x^2 y \right] dx,$$

and K is the class of functions $y(x)$ in $C_g^1[1, 2]$ satisfying $y(1) = y(2) = 0$. Then $p(x) = x$, $q(x) = (1 - x^2)/x$ and $f(x) = -x^2$ so the boundary value problem corresponding to minimizing $I[y]$ in K is an inhomogeneous Bessel equation

$$\frac{d}{dx}(xy') + \frac{x^2 - 1}{x}\, y = -x^2; \qquad y(1) = y(2) = 0.$$

(c)
$$I[y] = \int_1^2 [y'^2 - \lambda xy^2]\, dx,$$

and K is the class of functions $y(x)$ in $C_g^1[1, 2]$ satisfying $y(1) = y(2) = 0$. Here $p(x) \equiv 1$, $q(x) \equiv 0$ and $r(x) = x$ so the Euler-Lagrange equation is

$$y'' + \lambda xy = 0.$$

Since $r(x) > 0$ for $1 \le x \le 2$, the previous discussion implies that the first eigenfunction $y_1(x)$ is that function which minimizes

$$I_1[y] = \int_1^2 y'^2\, dx.$$

in the class of all functions in $C_g^1[1, 2]$ which satisfy $y(1) = y(2) = 0$ and $\int_1^2 xy^2(x)\, dx = 1$, and $I_1[y_1] = \lambda_1$.

Note that by making the transformation $z = -\lambda^{1/3} x$ in the differential equation one obtains the equation

$$\frac{d^2 y}{dz^2} - zy = 0.$$

Therefore the eigenfunctions are of the form

$$y_n(x) = a_n \, Ai(-\lambda_n^{1/3} x) + b_n \, Bi(-\lambda_n^{1/3} x),$$

where $Ai(z)$ and $Bi(z)$ are the Airy functions, λ_n is the nth eigenvalue, and a_n, b_n are appropriately chosen constants.

We conclude this section by making some remarks on the other types of boundary conditions obtainable for Equation (2) from the variational problem. First of all, in many cases we can require that the family K satisfy the boundary conditions $y(a) = y(b) = 0$ inasmuch as the change of variables

$$y(x) = z(x) + \frac{y_b - y_a}{b - a}(x - a) + y_a$$

will accomplish this.

Secondly if we do not impose any boundary conditions on K and suppose that $y_0(x)$ gives a minimum to the functional $I[y]$, then we obtain the "*natural*" *boundary conditions*

$$y'(a) = 0, \qquad y'(b) = 0$$

for Equation (2). To see this, let

$$\phi(t) = I[y_0 - t\eta]$$

$$= \int_a^b [p(x)(y_0'(x) + t\eta'(x))^2 + q(x)(y_0(x) + t\eta(x))^2$$

$$+ 2f(x)(y_0(x) + t\eta(x))] \, dx,$$

where $\eta(x)$ is any function in $C_g^1[a, b]$ and $|t|$ is small. As before we conclude that $\phi'(0) = 0$, which leads to the relation

$$-2 \int_a^b \left[\frac{d}{dx}(p(x)y_0'(x)) - q(x)y_0(x) - f(x) \right] \eta(x) \, dx$$

$$+ p(a)y_0'(a)\eta(a) - p(b)y_0'(b)\eta(b) = 0.$$

We assume that $p(x) > 0$, and since the above expression must be zero for *any* $\eta(x)$, it must be zero for any $\eta(x)$ satisfying $\eta(a) = \eta(b) = 0$, whence we obtain the Euler-Lagrange equation (2). Thus we may disregard the integral in the last expression, and now choosing any $\eta(x)$ that is zero at one endpoint and unity at the other, we obtain the natural boundary conditions. Note the further implication that we can require mixed boundary conditions of the type $y(a) = y'(b) = 0$ within the variational formulation.

Finally we consider the more general functional

$$I[y] = \int_a^b [p(x)y'^2 + q(x)y^2 + 2f(x)y] \, dx + \beta y^2(b) - \alpha y^2(a),$$

which depends also on the endpoint values of the function $y = y(x)$ chosen. By proceeding as in the derivation of the Euler equation one obtains the following boundary conditions:

$$p(a)y'(a) + \alpha y(a) = 0, \qquad p(b)y'(b) + \beta y(b) = 0.$$

These are the *homogeneous boundary conditions*; it is assumed that $p(x) > 0$ as before. Similar boundary conditions can be obtained for the Sturm-Liouville problem corresponding to Equation (4).

5.5   The Ritz Method

Now that we have shown the relation between the simplest problem in the calculus of variations and certain boundary problems, the next question is how can we use this relation to obtain approximations to the solution of the boundary value problem? One answer is to use the Ritz method which is a technique to approximate the greatest lower bound of a functional $I[y]$ in a class K of functions. We will first describe the method in the context of the calculus of variations then see how it is applied to the boundary value problem in the following section.

Let us suppose that we are given the functional

$$I[y] = \int_a^b f(x, y(x), x'(x))\, dx$$

and an admissible class K of functions, and we will assume that

$$i = \underset{y \in K}{\text{g.l.b. }} I[y] > -\infty$$

or equivalently that $I[y]$ is bounded below in K. Now by the definition of the greatest lower bound this implies that there is a sequence of functions $y_n(x)$, $n = 1, 2, \ldots$, in K such that $\lim_{n \to \infty} I[y_n] = i$. Such a sequence is called a *minimizing sequence*, and since in general we don't know how to find it we will need a little more machinery.

Therefore let us suppose that there is a denumerable family Φ of functions $\phi_n(x)$, $n = 1, 2, \ldots$, such that

(i)   finite linear combinations of elements of Φ belong to K, i.e. $\sum_1^n a_i \phi_i(x)$ is in K for any choice of the coefficients a_i and any positive integer n; and

(ii)   given any $\epsilon > 0$ and any element $y = y(x)$ in K there exists an element $\phi(x) = \sum_1^n a_i \phi_i(x)$ in Φ such that

$$|I[\phi] - I[y]| < \epsilon.$$

Note that we have *not* assumed that the family Φ is dense in K, that is, that we can approximate arbitrarily close any function in K with a finite linear combination of elements from Φ. When we come to the boundary value problem we will need this stronger assumption.

Now given any positive integer n let Φ_n be the set of all linear combinations $\{\sum_1^n a_i \phi_i(x)\}$. Then clearly Φ_n is a subset of K and we consider the function

$$F(a_1, a_2, \ldots, a_n) = I\left[\sum_1^n a_i \phi_i\right]$$

$$= \int_a^b f\left(x, \sum_1^n a_i \phi_i(x), \sum_1^n a_i \phi_i'(x)\right) dx,$$

which is merely the restriction of $I[y]$ to the set Φ_n. The function F is a real valued function defined on R^n and is bounded below by i—by suitable hypotheses on the integrand f we may assume F is continuously differentiable and therefore will have a minimum on any compact set of R^n. A necessary condition for a minimum is that it satisfy the relations

$$\frac{\partial F}{\partial a_i} = 0, \quad i = 1, 2, \ldots, n \tag{6}$$

so let us assume that $u_n(x) = \sum_1^n \bar{a}_i \phi_i(x)$ gives a minimum to $I[y]$ in the class Φ_n and hence the $\bar{a}_i$, $i = 1, 2, \ldots, n$ are solutions of Equations (6).

Since the family Φ_n is contained in the family Φ_{n+1} for any n, we have the relation $I[u_{n+1}] \leq I[u_n]$ since we are finding the minimum over a larger set of functions. Therefore the sequence $\{I[u_n]\}$ is bounded below by i, is monotone decreasing, and we assert that

$$\lim_{n \to \infty} I[u_n] = i.$$

To prove this, let $y_n(x)$, $n = 1, 2, \ldots$, be *any* minimizing sequence and therefore given any $\epsilon > 0$ there is an element $y_m(x)$ such that $I[y_m] < i + \epsilon$. By the assumption (ii) there is some positive integer r and an element $\phi(x)$ in Φ_r such that $I[\phi] < I[y_n] + \epsilon$. But the function $u_r(x)$ minimizes $I[y]$ in the family Φ_r, so we have the relation

$$i \leq I[u_r] \leq I[\phi] < I[y_n] + \epsilon < i + 2\epsilon,$$

which proves the assertion.

The rationale behind the Ritz method is that since the problem of finding a minimizing sequence and hence i is generally difficult, one reduces the problem to a finite dimensional one via Equations (6). Clearly the degree of approximation is completely dependent on the choice of the family Φ but in many cases the nature of $I[y]$ and the family K (*viz.*, the boundary conditions) will indicate a natural choice.

Finally note that the key assumption is (ii), which guarantees that the catastrophe

$$\lim_{n \to \infty} I[u_n] = i_0 > i$$

will not occur.

Example

We wish to find the minimum of

$$I[y] = \int_0^1 (1 + x)y'^2 \, dx$$

in the class of all functions $y = y(x)$ in $C_g^1 [0, 1]$ which satisfy $y(0) = 0$, $y(1) = 1$. If we made the transformation $y(x) = z(x) + x$, then the functional becomes

$$I_1[z] = \int_0^1 (1 + x)(z'^2 + 2z') \, dx + \tfrac{3}{2}$$

with boundary conditions $z(0) = z(1) = 0$. Then a natural choice for $\phi_n(x)$ would be $x^n(1 - x)$, and we would try approximations of the form

$$z_n(x) = \sum_1^n a_i \phi_i(x)$$

$$= a_1 x(1 - x) + a_2 x^2(1 - x) + \cdots + a_n x^n(1 - x).$$

Rather than this, let us use the first functional and approximations of the form

$$y_n(x) = x + \sum_1^n a_i \phi_i(x) = x + z_n(x)$$

with $z_n(x)$ as given above.

1st *Approximation*: $y_1(x) = x + a_1 x(1 - x)$. Then

$$I[y_1] = \int_0^1 (1 + x)y_1'(x)^2 \, dx = \frac{3}{2} - \frac{a_1}{3} - \frac{a_1^2}{2} = F(a_1),$$

and the relation $\partial F/\partial a_1 = 0$ implies $a_1 = \tfrac{1}{3}$. Therefore

$$u_1(x) = x + \tfrac{1}{3}x(1 - x) \qquad \text{and} \qquad I[u_1] = 1.4444 \cdots.$$

2nd *Approximation*:

$$y_2(x) = x + a_1 x(1 - x) + a_2 x^2(1 - x).$$

Then

$$I[y_2] = \int_0^1 (1 + x)y_2'(x)^2 \, dx$$

$$= \tfrac{3}{2} - \tfrac{1}{2}a_1^2 + \tfrac{7}{30}a_2^2 - \tfrac{1}{3}a_1 + \tfrac{17}{30}a_1 a_2$$

$$= F(a_1, a_2),$$

and the relations $\partial F/\partial a_1 = 0$, $\partial F/\partial a_2 = 0$ give that $a_1 = 55/131$, $a_2 = -20/131$. Therefore

$$u_2(x) = x + \tfrac{55}{131} x(1 - x) - \tfrac{20}{131} x^2(1 - x)$$

and

$$I[u_2] = 1.4427 \cdots.$$

Since $p(x) = 1 + x$ is positive on $0 \leq x \leq 1$, the result of the previous section applies and we know that the solution of the Euler-Lagrange equation

$$\frac{d}{dx}((1 + x)y') = 0; \quad y(0) = 0, \quad y(1) = 1,$$

gives an absolute minimum to $I[y]$ in K. The solution is

$$y_0(x) = \frac{\log(1 + x)}{\log 2} \quad \text{and} \quad I[y] = 1.4427 \cdots,$$

so we obtained two-place accuracy with just the first approximation. The following table indicates that in this case the Ritz method provided an excellent approximation of the minimizing function as well.

$$x = \tfrac{1}{4}: \quad y_0(x) = 0.3218 \quad u_2(x) = 0.3215,$$

$$x = \tfrac{1}{2}: \quad y_0(x) = 0.5849 \quad u_2(x) = 0.5859,$$

$$x = \tfrac{3}{4}: \quad y_0(x) = 0.8073 \quad y_2(x) = 0.8093$$

The particular approximations of i and $y_0(x)$ obtained above were certainly the result of an appropriate choice of the functions $\phi_n(x)$. If we examine the functional $I_1[z]$ with boundary conditions $z(0) = z(1)$ on the class K, then for any function $z(x)$ in K we can find a function

$$\phi(x) = \sum_1^n a_i \phi_i(x) = (1 - x)(a_1 x + a_2 x^2 + \cdots + a_n x^n)$$

in Φ such that for $0 \leq x \leq 1$ the quantities $|z(x) - \phi(x)|$, $|z'(x) - \phi'(x)|$ can be made uniformly small. This is a consequence of the fact that the family Φ is a linearly independent set of functions which are dense in K by the Weierstrass approximation theorem.

In turn this implies that the difference $|I_1[\phi] - I_1[z]|$ will be small and assumption (ii) will be satisfied. Therefore the Ritz method will produce a minimizing sequence and as a side benefit we obtain a good approximation to the minimizing function. This leads us to an alternate assumption for (ii) in the case where the family Φ is dense in K:

(ii)′   Given any $\epsilon > 0$ and any element $y = y(x)$ in K, there exists a $\delta > 0$ and a function $\phi(x) = \sum_1^n a_i \phi_i(x)$ in Φ such that $|\phi(x) - y(x)| < \delta$, $|\phi'(x) - y'(x)| < \delta$ for $0 \le x \le 1$, and this in turn implies that $|I[\phi] - I[y]| < \epsilon$.

5.6   Application to the Boundary Value Problem

We will now apply the Ritz method and some of the results of the previous sections to the problem of approximating the solutions or eigenvalues and eigenfunctions of boundary value problems. To begin with we consider the equation

$$\frac{d}{dx}(p(x)y') - q(x)y - f(x) = 0, \tag{7}$$

and for ease of exposition and calculation we will suppose the boundary conditions are

$$y(0) = y(l) = 0.$$

Other types of boundary conditions will be remarked upon later in the discussion.

The functional corresponding to Equation (7) is

$$I[y] = \int_0^l [p(x)y'^2 + q(x)y^2 + 2f(x)y]\, dx,$$

and natural choices for the family Φ in the Ritz method are

$$\phi_n(x) = x^n(l - x) \quad \text{or} \quad \phi_n(x) = \sin\frac{n\pi x}{l}, \quad n = 1, 2, \ldots.$$

These have the advantage that they are a linearly independent family and are dense in the sense described in the previous section. Finally, we make the usual assumptions $(p(x) > 0,\ q(x) \ge 0)$ to insure that the solution of (7) gives a unique absolute minimum to $I[y]$.

Proceeding as in the Ritz method we form the nth approximation $y_n(x) = \sum_1^n a_i \phi_i(x)$, and after a little calculation we obtain

$$I[y_n] = F(a_1, a_2, \ldots, a_n)$$
$$= \sum_{i,\, j=1}^n \alpha_{ij} a_i a_j + 2\sum_1^n \beta_i a_i,$$

where

$$\alpha_{ij} = \alpha_{ji} = \int_0^l [p(x)\phi_i'(x)\phi_j'(x) + q(x)\phi_i(x)\phi_j(x)]\, dx$$

and

$$\beta_i = \int_0^l f(x)\phi_i(x)\, dx.$$

The relations $\partial F/\partial a_i = 0$, $i = 1, 2, \ldots, n$ lead to a system of equations (after dividing by 2)

$$\sum_{j=1}^n \alpha_{ij} a_j + \beta_i = 0, \quad i = 1, 2, \ldots, n, \tag{8}$$

which is n linear equations in the n unknowns $a_1, a_2, \ldots, a_n$. Let $a_i = \bar{a}_i$, $i = 1, 2, \ldots, n$ be the solution of the system (8) and we assert that:

The solution of the system (8) is unique.

If it were not, then the homogeneous system $\sum_{j=1}^n \alpha_{ij} a_j = 0$ would have a nontrivial solution $a_i = \hat{a}_i$, $i = 1, 2, \ldots, n$, and from the definition of the α_{ij} this would give the relation

$$\int_0^l [p(x)\hat{y}_n'(x)\phi_i'(x) + q(x)\hat{y}_n(x)\phi_i(x)]\, dx = 0, \quad i = 1, 2, \ldots, n,$$

where $\hat{y}_n(x) = \sum_1^n \hat{a}_j \phi_j(x)$. Multiply the last equation by $\hat{a}_i$ and sum from 1 to n to obtain

$$\int_0^l [p(x)\hat{y}_n'(x)^2 + q(x)\hat{y}_n(x)^2]\, dx = 0,$$

which can only occur if $\hat{y}_n(x) \equiv 0$. Since all the $\hat{a}_i$ were not zero, this contradicts the linear independence of the family Φ.

We can now construct the minimizing sequence $\sum_1^n \bar{a}_i \phi_i(x)$, $n = 1, 2, \ldots$, and $\lim_{n \to \infty} I[u_n] = \iota$, the minimum of the functional. But furthermore we can assert that

The minimizing sequence $u_n(x)$, $n = 1, 2, \ldots$, constructed by the Ritz method converges uniformly to the solution of Equation (7).

The solution $y_0(x)$ of (7) gives an absolute minimum to $I[y]$ and if we compute the difference $I[u_n] - I[y_0]$, we obtain (see Section 5.4),

$$I[u_n] - I[y_0] = \int_0^l [p(x)(u_n'(x) - y_0'(x))^2 + q(x)(u_n(x) - y_0(x))^2]\, dx.$$

If p_0 is the minimum of $p(x)$ on $0 \leq x \leq l$, then $p_0 > 0$ and the last expression is less than or equal to

$$p_0 \int_0^l [u_n'(x) - y_0'(x)]^2 \, dx,$$

and we obtain the estimate

$$\int_0^l [u_n'(x) - y_0'(x)]^2 \, dx \leq \frac{1}{p_0} (I[u_n] - I[y_0]).$$

Now using the Cauchy-Schwarz-Bunyakovsky inequality we have that for $0 \leq x \leq l$,

$$|u_n(x) - y_0(x)| = \left| \int_0^x [u_n'(t) - y_0'(t)] \, dt \right|$$

$$\leq \left[\int_0^x [u_n'(t) - y_0'(t)]^2 \, dt \int_0^x dt \right]^{1/2}$$

$$\leq \frac{l^{1/2}}{\sqrt{p_0}} \sqrt{I[u_n] - I[y_0]}.$$

But the last quantity approaches zero as $n \to \infty$ and is independent of x, so we conclude

$$\lim_{n \to \infty} u_n(x) = y_0(x)$$

uniformly on $0 \leq x \leq l$.

Example

To find an approximation to the solution of

$$y'' + (1 + x)y = 1; \qquad y(0) = y(1) = 0.$$

The corresponding functional is

$$I[y] = \int_0^1 [y'^2 + (1 + x)y^2 + 2y] \, dx$$

and the two natural choices for the $\phi_n(x)$ are

$$\phi_n(x) = x^n(1 - x) \qquad \text{or} \qquad \phi_n(x) = \sin n\pi x.$$

Before jumping in with pencil, tables and slide rule, let us try to develop some rationale for which of the two approximations to use.

If we try an approximation of the form

$$y_n(x) = \sum_1^n a_j \sin j\pi x$$

and substitute it in the left-hand side of the differential equation we will clearly obtain an expression which vanishes at $x = 0$ and $x = 1$. Since the right-hand side is equal to 1, the approximate solution $y_n(x)$ will be a worthless fit to the actual solution near the endpoints of the interval. Therefore, the choice $\phi_n(x) = x^n(1 - x)$ is clearly the better one.

1st *Approximation*: $y_1(x) = a_1 x(1 - x) = a_1 \phi_1(x)$.

Since $p(x) = 1$, $q(x) = 1 + x$, and $f(x) = 1$, we have

$$\alpha_{11} = \int_0^1 [\phi_1'(x)^2 + (1 + x)\phi_1(x)^2] \, dx$$

$$= \int_0^1 [(1 - 2x)^2 + (1 - x)x^2(1 - x)^2] \, dx = \tfrac{23}{60};$$

$$\beta_1 = \int_0^1 \phi_1(x) \, dx = \int_0^1 x(1 - x) \, dx = \tfrac{1}{6};$$

and Equation (8) is

$$\alpha_{11} a_1 + \beta_1 = \tfrac{23}{60} a_1 + \tfrac{1}{6} = 0,$$

and $a_1 = \tfrac{-10}{23} = -0.435 \cdots$. Therefore the first approximation is $y_1(x) = -0.435x(1 - x)$.

2nd *Approximation*:

$$y_2(x) = a_1 x(1 - x) + a_2 x^2(1 - x) = a_1\phi_1(x) + a_2 \phi_2(x).$$

Then α_{11} and β_1 are as before, and

$$\alpha_{22} = \int_0^1 [\phi_2'(x)^2 + (1 + x)\phi_2(x)^2] \, dx$$

$$= \int_0^1 [(2x - 3x^2)^2 + (1 + x)x^4(1 - x)^2] \, dx = \tfrac{25}{168};$$

$$\alpha_{12} = \alpha_{21} = \int_0^1 [\phi_1'(x)\phi_2'(x) + (1 + x)\phi_1(x)\phi_2(x)] \, dx$$

$$= \int_0^1 [(1 - 2x)(2x - 3x^2) + (1 + x)x^3(1 - x)^2] \, dx$$

$$= \tfrac{27}{140};$$

$$\beta_2 = \int_0^1 \phi_2(x) \, dx = \int_0^1 x^2(1 - x) \, dx = \tfrac{1}{12},$$

so Equations (8) are

$$\frac{27}{60} a_1 + \frac{27}{140} a_2 + \frac{1}{6} = 0$$

$$\frac{27}{140} a_1 + \frac{25}{168} a_2 + \frac{1}{12} = 0.$$

The solutions are $a_1 = -0.435 \cdots$, $a_2 = 0.0111 \cdots$, so the second approximate solution is

$$y_2(x) = -0.435x(1 - x) + 0.0111x^2(1 - x).$$

The reader may wish to verify that if one uses the approximation $\tilde{y}_1(x) = a_1 \sin \pi x$ in the Ritz method, one obtains that $a_1 = -0.112$. If we compare the two first approximations we see that

$$y_1'' + (1 + x)y_1 = 0.870 - 0.435x + 0.435x^2,$$

$$\tilde{y}_1'' + (1 + x)\tilde{y}_1 = 0.993 \sin \pi x - 0.112x \sin \pi x,$$

and the graphs indicate the degree of approximation to 1.

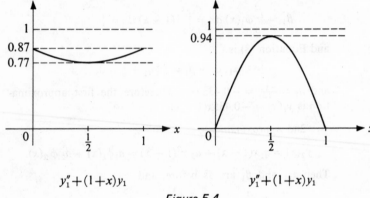

$$y_1'' + (1+x)y_1 \qquad\qquad y_1'' + (1+x)y_1$$

Figure 5.4

In the case of other types of boundary conditions some possible choices for the $\phi_n(x)$ are given below:

(a) $$y'(0) = y'(l) = 0,$$

$$\phi_0(x) = 1, \quad \phi_1(x) = 2x^3 - 3lx^2, \quad \phi_k(x) = x^k(l - x)^2, \quad k = 2, 3, \ldots,$$

or

$$\phi_0(x) = 1, \quad \phi_k = \cos \frac{k\pi x}{l}, \quad k = 1, 2, \ldots,$$

(b) $$y'(0) + \alpha y(0) = 0, \quad y'(l) + \beta y(l) = 0,$$

$$\phi_0(x) = x^2\left(x - l - \frac{l}{2 + \beta l}\right), \quad \phi_1(x) = (l - x)^2\left(x + \frac{1}{2 - \alpha l}\right)$$

$$\phi_k(x) = x^k(l - x)^2, \quad k = 2, 3, \ldots.$$

As the above example indicates, the degree of the approximation greatly depends on the choice of the $\phi_n(x)$. While in many problems the choices above, or suitable modifications will usually suffice, certain problems may call for other choices such as orthogonal polynomials (Legendre, Laguerre, or Hermite polynomials) or special functions such as the Bessel functions. Some investigation of the problem on hand should be made before proceeding to do the approximations.

The *most important* use of the Ritz method for ordinary differential equations is to approximate the eigenvalues and possibly the eigenfunctions of a boundary value problem of the Sturm-Liouville type. Rather than go into a detailed description we will illustrate its use by some examples. The reason for this is that we will discuss this problem again in the next section when we consider Galerkin's method, which agrees with the Ritz method in many cases.

Suppose we are given the eigenvalue problem

$$\frac{d}{dx}(p(x)y') - q(x)y + \lambda r(x)y = 0; \qquad y(0) = y(l) = 0,$$

where we assume that $p(x)$ and $r(x)$ are positive on $0 \leq x \leq l$. From our discussion in Section 5.4 we know that the first eigenvalue λ_1 is the minimum of the functional

$$I[y] = \int_0^l [p(x)y'^2 + q(x)y^2] \, dx$$

in the class of all functions in $C_g^1[0, l]$ satisfying

$$\int_0^1 r(x)y^2(x) \, dx = 1; \qquad y(0) = y(l) = 0.$$

Suppose that Φ is a linearly independent family of functions $\phi_n(x)$ which satisfy the boundary conditions and are dense (in the sense of (ii′) of Section 5.5). Consider a second approximation of the form

$$y_2(x) = a_1\phi_1(x) + a_2\phi_2(x).$$

Then proceeding as in the Ritz method, we can compute

$$I[y_2] = \int_0^l [p(x)y_2'(x)^2 + q(x)y_2(x)^2] \, dx$$

$$= \alpha_{11}a_1^2 + \alpha_{12}a_1a_2 + \alpha_{21}a_1a_2 + \alpha_{22}a_2^2 = F(a_1, a_2),$$

where the α_{ij} are computed as before. Furthermore $y_2(x)$ must satisfy the constraint

$$\int_0^l r(x)y_2(x)^2 \, dx = \gamma_{11}a_1^2 + \gamma_{12}a_1a_2 + \gamma_{21}a_1a_2 + \gamma_{22}a_2^2$$

$$= R(a_1, a_2) = 1,$$

where

$$\gamma_{ij} = \gamma_{ji} = \int_0^l r(x)\phi_i(x)\phi_j(x) \, dx.$$

The finite dimensional problem is then to minimize the function $F(a_1, a_2)$ subject to the constraint $R(a_1, a_2) = 1$.

This problem is solved using Lagrange multipliers—namely one considers the function

$$G(a_1, a_2) = F(a_1, a_2) + \lambda(1 - R(a_1, a_2))$$

and a necessary condition for a minimum is that

$$\frac{\partial G}{\partial a_1} = 0, \quad \frac{\partial G}{\partial a_2} = 0, \quad \frac{\partial G}{\partial \lambda} = 0.$$

The choice of the symbol λ to denote the multiplier is not a notational blunder as the reader will see.

The last equations lead to the following set of equations in the three unknowns a_1, a_2 and λ:

$$\begin{cases} (\alpha_{11} - \lambda\gamma_{11})a_1 + (\alpha_{12} - \lambda\gamma_{12})a_2 = 0 \\ (\alpha_{21} - \lambda\gamma_{21})a_1 + (\alpha_{22} - \lambda\gamma_{22})a_2 = 0 \end{cases}; \quad R(a_1, a_2) = 1. \quad (9)$$

The bracketed pair of equations will have a nontrivial solution a_1, a_2 only if the determinant

$$\Delta = \det\begin{pmatrix} \alpha_{11} - \lambda\gamma_{11} & \alpha_{12} - \lambda\gamma_{12} \\ \alpha_{21} - \lambda\gamma_{21} & \alpha_{22} - \lambda\gamma_{22} \end{pmatrix} = 0.$$

This determinant is a second degree polynomial in λ and its smallest root will be an approximation $\lambda_1^{(2)}$ of the smallest eigenvalue λ_1. If we substitute the value $\lambda_1^{(2)}$ in the first pair of equations of (9) and use the second equation, we can find the approximate values $\bar{a}_1$ and $\bar{a}_2$. Then the approximation to the first eigenfunction is

$$y_1^{(2)}(x) = \bar{a}_1\phi_1(x) + \bar{a}_2\phi_2(x).$$

The larger root of Δ will actually be a (generally poor) approximation $\lambda_2^{(2)}$ of the second eigenvalue λ_2 and we can proceed as before to find an approximation $y_2^{(2)}(x)$ to the second eigenfunction. In general, if we used an approximation of the form $y_n(x) = \sum_1^n a_i\phi_i(x)$, then the corresponding determinant Δ would be a polynomial of degree n in λ. Its n roots, ordered by magnitude, would be approximations to the first n eigenvalues and the accuracy would drop off as the roots became larger.

Example

Consider the simple problem

$$y'' + \lambda y = 0; \quad y(0) = y(1) = 0,$$

whose eigenvalues are $\lambda_n = n^2\pi^2$ and corresponding normalized eigenfunctions are $\sqrt{2}\sin n\pi x$, $n = 1, 2, \ldots$. Then $p(x) = r(x) \equiv 1$ and $q(x) = 0$ so if we let $\phi_n(x) = x^n(1 - x)$, then we obtain the following equations corresponding to (9):

$$\left.\begin{array}{l} \left(\dfrac{1}{3} - \dfrac{\lambda}{30}\right)a_1 + \left(\dfrac{1}{6} - \dfrac{\lambda}{60}\right)a_2 = 0 \\[2mm] \left(\dfrac{1}{6} - \dfrac{\lambda}{60}\right)a_1 + \left(\dfrac{2}{15} - \dfrac{\lambda}{105}\right)a_2 = 0 \end{array}\right\};$$

$$\frac{1}{30}a_1^2 + \frac{1}{30}a_1a_2 + \frac{1}{105}a_2^2 = 1.$$

Therefore

$$\Delta = \det\begin{pmatrix} \dfrac{1}{3} - \dfrac{\lambda}{30} & \dfrac{1}{6} - \dfrac{\lambda}{60} \\[3mm] \dfrac{1}{6} - \dfrac{\lambda}{60} & \dfrac{2}{15} - \dfrac{\lambda}{105} \end{pmatrix} = \frac{(10 - \lambda)(42 - \lambda)}{60 \cdot 420},$$

and its roots are $\lambda_1^{(2)} = 10$, $\lambda_2^{(2)} = 42$. The actual eigenvalues are $\lambda_1 = \pi^2 = 9.8696\cdots$ and $\lambda_2 = 4\pi^2 = 39.477\cdots$, so we do obtain a good approximation of λ_1 and only a fair one of λ_2.

Substituting $\lambda = 10$ in the above equations leads to the solution $a_2 = 0$ and the second relation becomes $\frac{1}{30}a_1^2 = 1$. Therefore an approximation to the first eigenfunction $\sqrt{2}\sin \pi x$ is $y_1^{(2)} = \sqrt{30}x(1 - x)$. In the same manner, substituting $\lambda = 42$ gives that $a_2 = -2a_1$ and $a_1^2 = 210$, therefore the approximation to $\sqrt{2}\sin 2\pi x$ is

$$y_2^{(2)}(x) = \sqrt{210}\,[x(1 - x) - 2x^2(1 - x)].$$

A comparison of $\sqrt{2}\sin \pi x$ and $y_1^{(2)}(x)$ will show that a reasonably good match is obtained for a rough approximation.

Finally we remark that the Ritz method is applicable in some cases where the hypotheses $p(x) > 0$ or $r(x) > 0$ in the eigenvalue problem are not satisfied in the sense that $p(x)$ or $r(x)$ vanishes at one or both of the endpoints of the interval. The reader should browse through some of the references given to see other applications of Ritz's method and further discussion of it.

5.7 The Galerkin Method

To conclude this chapter we will briefly describe the Galerkin method, which in many cases gives the same determining equations as the Ritz method. However, it is more widely applicable because it can be used in cases where the differential equation is not the Euler-Lagrange equation of a functional, i.e., is not related to a variational problem.

To introduce the method let us go back to the determining Equations (8) of the previous section, which we write in expanded form

$$\sum_{j=1}^{n} \left(\int_0^l [p(x)\phi_i'(x)\phi_j'(x) + q(x)\phi_i(x)\phi_j(x)] \, dx \right) a_j$$
$$+ \int_0^l f(x)\phi_i(x) \, dx = 0, \quad i = 1, \ldots, n.$$

Recalling that $y_n(x) = \sum_1^n a_j \phi_j(x)$, we can write instead

$$\int_0^l [p(x)\phi_i'(x)y_n(x) + (q(x)y_n(x) + f(x))\phi_i(x)] \, dx = 0$$

$i = 1, \ldots, n$, and since we assumed that $\phi_i(0) = \phi_i(l) = 0$, we can integrate the first term by parts to finally obtain

$$\int_0^l \left[-\frac{d}{dx}(p(x)y_n(x)) + q(x)y_n(x) + f(x) \right] \phi_i(x) \, dx = 0$$

$i = 1, 2, \ldots, n$. If we write the differential equation as

$$L(y) = \frac{d}{dx}(p(x)y') - q(x)y - f(x) = 0,$$

then we see that the previous equations can more easily be written as

$$\int_0^l L(y_n(x))\phi_i(x) \, dx = 0, \quad i = 1, 2, \ldots, n;$$
$$y_n(x) = \sum_1^n a_i \phi_i(x).$$

These are a form of the Galerkin equations and in this case we can immediately see one advantage of using them, namely, that of convenience of notation.

More generally, suppose we are given an ordinary differential equation (not necessarily linear), which we can denote by $L(y) = 0$, and boundary conditions to be satisfied at $x = a$ and $x = b$. Let $Y_n = \{\theta(x, a_1, \ldots, a_n)\}$ be a family of approximating functions, dependent on the parameters $a_1, \ldots, a_n$ and satisfying the boundary conditions, and let $\Phi = \{\phi_n(x)\}$ be a family of linearly independent functions on $a \leq x \leq b$.

We seek a solution of the form $y_n(x) = \theta(x, \bar{a}_1, \ldots, \bar{a}_n)$, i.e., a member of the family Y_n, where the values $\bar{a}_i$ of the parameters are determined by solving the orthogonality conditions

$$\int_a^b L(y_n)\phi_i(x)\,dx = \int_a^b L(\theta(x, a_1, \ldots, a_n))\phi_i(x)\,dx = 0, \quad i = 1, \ldots, n.$$

These are the *Galerkin equations* and the most common case is when the boundary conditions are homogeneous, the family Y_n is $\Phi_n = \{\sum_1^n a_i \phi_i(x)\}$ and hence we look for an approximation of the form $y_n(x) = \sum_1^n \bar{a}_j \phi_j(x)$. Then the Galerkin equations are

$$\int_a^b L(y_n)\phi_i(x)\,dx = \int_a^b L\left(\sum_1^n a_j \phi_j(x)\right)\phi_i(x)\,dx = 0, \quad i = 1, \ldots, n,$$

and these must be solved for the values $\bar{a}_i$; clearly the equations are linear in the a_i if the differential equation is a linear one.

Example

We wish to approximate the solution of the boundary value problem

$$xy'' + y' - (x - 3)y = 0; \quad y(0) = 1, \quad y(\tfrac{1}{2}) = 0.$$

First of all make the transformation $y(x) = z(x) + (1 - 2x)$, and the equation becomes

$$L(z) = xz'' + z' - (x - 3)z + 2x^2 - 7x + 1 = 0;$$

$$z(0) = z(\tfrac{1}{2}) = 0.$$

We let $\Phi = \{x^n(\tfrac{1}{2} - x)\}$, and since the boundary conditions are homogeneous we will try approximations of the form $z_n(x) = \sum_1^n a_j x^j(\tfrac{1}{2} - x)$. For a first approximation $z_1(x) = a_1 x(\tfrac{1}{2} - x)$ the Galerkin equation is then

$$\int_0^{\frac{1}{2}} L(z_1(x))x(\tfrac{1}{2} - x)\,dx = 0$$

or

$$\int_0^{\frac{1}{2}} [a_1(\tfrac{1}{2} - \tfrac{5}{2}x - \tfrac{7}{2}x^2 + x^3) + (2x^2 - 7x + 1)]x(\tfrac{1}{2} - x)\,dx = 0.$$

The reader should verify that this gives $a_1 = -\tfrac{96}{29}$, hence

$$y_1(x) = -\tfrac{96}{29}x(\tfrac{1}{2} - x) + (1 - 2x) = 1 - \tfrac{106}{29}x + \tfrac{96}{29}x^2.$$

This should be compared with the actual solution $y(x) = (1 - 2x)e^{-x}$.

Given the eigenvalue problem

$$L(y) = a(x)y'' + b(x)y' + c(x)y + \lambda r(x)y = 0$$

with homogeneous boundary conditions at $x = a$ and $x = b$, suppose $\Phi = \{\phi_n(x)\}$ are a family of linearly independent functions satisfying the boundary conditions. The Galerkin equations for the nth approximation $y_n(x) = \sum_1^n a_j \phi_j(x)$ are then

$$\int_a^b L(y_n(x))\phi_i(x)\,dx = \sum_{j=1}^n (\alpha_{ij} + \lambda\gamma_{ij})a_j = 0, \quad i = 1, \ldots\ n,$$

where

$$\alpha_{ij} = \int_a^b [a(x)\phi_i(x)\phi_j''(x) + b(x)\phi_i(x)\phi_j'(x) + c(x)\phi_i(x)\phi_j]\,dx$$

and

$$\gamma_{ij} = \int_a^b r(x)\phi_i(x)\phi_j(x)\,dx.$$

As in the Ritz method one sees that the system of equations in the a_i will have a solution only if the determinant

$$\Delta = \det\begin{pmatrix} \alpha_{11} + \lambda\gamma_{11} & \cdots & \alpha_{1n} + \lambda\gamma_{1n} \\ \vdots & & \vdots \\ \alpha_{n1} + \lambda\gamma_{n1} & \cdots & \alpha_{nn} + \lambda\gamma_{nn} \end{pmatrix}$$

equals zero. This leads to a polynomial of degree n in λ; its roots will be approximations to the eigenvalues. By substituting the value of a root in the system of equations one can then determine the corresponding approximate eigenfunction (up to a multiplicative constant).

Example

We consider the eigenvalue problem

$$L(y) = y'' + \lambda xy = 0; \qquad y(0) = y(\pi) = 0,$$

and let $\Phi = \{\sin nx\}$. If we try an approximation $y_2(x) = a_1 \sin x + a_2 \sin 2x$, the Galerkin equations are

$$\int_0^\pi L(y_2)\sin x\,dx$$

$$= \int_0^\pi [a_1(\lambda x - 1)\sin x + a_2(\lambda x - 4)\sin 2x]\sin x\,dx = 0,$$

$$\int_0^\pi L(y_2)\sin 2x\,dx$$

$$= \int_0^\pi [a_1(\lambda x - 1)\sin x + a_2(\lambda x - 4)\sin 2x]\sin 2x\,dx = 0.$$

Performing the integration, one obtains the equations

$$\left(\frac{\pi}{2} - \lambda \frac{\pi^2}{4}\right) a_1 + \frac{17\lambda}{18} a_2 = 0,$$

$$\frac{17\lambda}{18} a_1 + \left(2\pi - \lambda \frac{\pi^2}{16}\right) a_2 = 0$$

and $\Delta = 0.630\lambda^2 - 16.47\lambda + 9.87$, whose roots are $\lambda_1^{(2)} = 0.611$ and $\lambda_2^{(2)} = 25.53$. Substituting say $\lambda_1^{(2)}$ in the above system of equations gives the relation $a_2 \cong -\frac{1}{10} a_1$, so the first approximate eigenfunction is

$$y_1^{(2)} = a_1[\sin x - \tfrac{1}{10} \sin 2x].$$

The reader will note (see Example c, Section 5.4) that the solutions are of the form

$$y(x) = a_1 Ai(-\lambda^{1/3}x) + b_1 Bi(-\lambda^{1/3}x),$$

where $Ai(x)$ and $Bi(x)$ are the Airy functions. The relation $y(0) = 0$ implies one must find those values of $\lambda > 0$ for which $Ai(-\lambda^{1/3}\pi) - 0.577Bi(-\lambda^{1/3}\pi) = 0$. From a table one can obtain the first value $\lambda_1 = 0.608$ so the approximation $\lambda_1^{(2)}$ is an excellent one.

We conclude this chapter by noting that the Galerkin method has recently become of great interest in the proofs of existence of periodic solutions of certain nonlinear differential equations and in their approximation. The following example illustrates the approximation technique.

Example

Given the nonlinear differential equation

$$L(x) = \ddot{x} + x^3 - \sin t = 0,$$

we seek an odd 2π-periodic solution satisfying $x(\pi/2 + t) = x(\pi/2 - t)$. Therefore let $\Phi = \{\sin(2n - 1)t\}$ and if the nth approximation is

$$x_n(t) = b_1 \sin t + b_3 \sin 3t + \cdots + b_{2n-1} \sin(2n - 1)t,$$

then the corresponding Galerkin equations are

$$\int_0^{2\pi} L(x_n(t))\sin(2k - 1)t \, dt = 0, \quad k = 1, 2, \ldots, n.$$

For instance if $n = 1$, then $x_1(t) = b_1 \sin t$ and the Galerkin equation is

$$\int_0^{2\pi} L(x_1(t)) \sin t \, dt$$

$$= \int_0^{2\pi} [-b_1 \sin t + b_1^3 \sin^3 t - \sin t] \sin t \, dt$$

$$= 2\left(-\frac{\pi}{2} b_1 + \frac{3\pi}{8} b_1^3 - \frac{\pi}{2}\right) = 0,$$

which has the solution $b_1 = 1.49 \cdots$, hence the first approximation is $x_1(t) = (1.49 \cdots) \sin t$.

In the example given the fundamental questions are

(i) whether the given equation has a periodic solution; and
(ii) if so, whether the Galerkin approximations are close to the periodic solution.

These are very deep mathematical questions in the area of functional analysis and the reader is referred to the papers of Professor L. Cesari, the principal investigator of this approach to nonlinear problems.

Problems

1. Show that the functional

$$I[y] = \int_0^1 xy'^2 \, dx$$

has no minimum in the class K of functions in $C_s^1[0, 1]$ satisfying $y(0) = 1$, $y(1) = 0$. Do this by
(i) showing that $I[y] > 0$ for every y in K, hence $\dot{i} = $ g.l.b. $I[y] \geq 0$;
(ii) constructing a minimizing sequence $\{y_n(x)\}$ in K satisfying

$$\lim_{n \to \infty} I[y_n] = 0.$$

Hence, $\dot{i} = 0$ so the minimum cannot be attained. (*Hint:* Try functions of the form $y_n(x) = 1$ if $0 \leq x \leq 1/n$, $y_n(x) = \theta_n(x)$ if $1/n \leq x \leq 1$ where $\theta_n'(x)^2 \to 0$ as $n \to \infty$ if $x > 1/n$ but $\theta'(1/n)^2 \to \infty$.)

2a. From the Euler-Lagrange equation obtain the following form of the equation for integrands of the form $f = f(y, y')$;

$$f - y'f_{y'} = c, \quad \text{a constant}.$$

2b. Use the above form to show that a possible solution for the minimal surface of revolution with

$$I[y] = \int_a^b y \sqrt{1 + y'^2} \, dx$$

is a catenary.

3. Show that there are infinitely many functions which give a minimum to the functional

$$I[y] = \int_0^2 y'^2(1 + y')^2 \, dx$$

in the class K of all functions in $C_o^1[0, 2]$ satisfying $y(0) = 1$, $y(2) = 0$.

4a. Given the functional

$$I[y] = \int_b^a p^2(x)y'^2 \, dx,$$

show that its absolute minimum in the class K of all functions in $C_o^1[a, b]$ satisfying $y(a) = y_a$, $y(b) = y_b$ is

$$i = (y_b - y_a)^2 \left(\int_a^b p(x)^{-2} \, dx \right)^{-1}$$

(*Hint:* Use the Cauchy-Schwarz-Bunyakovsky inequality.)

4b. Use the above result to obtain a lower bound for i for the functional

$$I[y] = \int_0^1 [(x^2 + 1)y'^2 + e^{-2x}y^2]dx; \quad y(0) = 0, \quad y(1) = 2,$$

then use the Ritz method to obtain an upper bound.

5. Use the Ritz method to obtain an approximation to the minimum i of the following functionals in the class K of functions in $C_o^1[0, 1]$ satisfying the boundary conditions given.

(i) $$I[y] = \int_0^1 (y'^2 + y^2)dx; \quad y(0) = 1, \quad y(1) = 3.$$

Here the minimum is given by the solution $y_0(x)$ of the Euler-Lagrange equation. Find it, compute $I[y_0]$, and compare your answer and the degree of approximation to $y_0(x)$.

(ii) $$I[y] = \int_0^1 [y'^2 + y^2 + 2xy]dx; \quad y(0) = y(1) = 0.$$

Do the same as in (i).

(iii) $$I[y] = \int_0^1 [y'^2 + x^2y^2]dx; \quad y(0) = y'(1) = 0.$$

(iv) $$I[y] = 2y(1)^2 + \int_0^1 [(x^2 + 2)y'^2 + y^2]dx; y(0) = 0.$$

6. Using the Ritz or Galerkin method with polynomial approximations, estimate the first two eigenvalues of the equation $y'' + \lambda y = 0$ under

the given boundary conditions. Compare the values you obtain with the actual values:

(i) $y(-1) = y(1) = 0,$

(ii) $y(0) + y'(1) = 0,$

(iii) $y(0) = 0, \quad y'(1) + y(1) = 0,$

(iv) $y'(0) = y'(\pi) = 0.$

7.   Use the Ritz or Galerkin method to approximate the first two eigenvalues of the given boundary value problem. For those cases marked with an asterisk compare your answers with the actual values obtained by solving the equation.

(i)* $x^2 y'' + \lambda y = 0; \; y(1) = y(e^{\pi}) = 0.$

(ii) $y'' + \lambda(1 + x^2)y = 0; \; y(0) = y'(1) = 0 \;\; (\lambda_1 < 1.601, \lambda_2 < 18.2).$

(iii) $\dfrac{d}{dx} (\sqrt{1 + x}\, y') + \lambda y = 0; \; y(0) = y(1) = 0 \;\; (\lambda_1 \cong 12.005).$

(iv)* $\dfrac{d}{dx} (xy') + \lambda xy = 0; \; y(1) = 0, |y(0)| < \infty.$ (Use $\phi_1(x) = 1 - x^2,$

        $\phi_2(x) = x^2(1 - x^2)$.)

(v) $\dfrac{d}{dx} ((1 + x)y') + \lambda(1 + x)y = 0; \; y'(0) = y(1) = 0.$

       (Try $\phi_n(x) = 1 - x^{n+1}$.)

(vi)* $y'' + \lambda xy = 0; \quad y(0) = y(1) = 0.$

(vii) $x^4 y'' + \lambda y = 0; \quad y(1) = y(2) = 0 \;\; (\lambda_n = 4\pi^2 n^2).$

(viii)* $\dfrac{d}{dx} (xy') + \lambda x^{-1} y = 0, \; y(1) = y(e^{\pi}) = 0.$

(ix)* $y'' + \lambda r(x)y = 0; \quad y(0) = y(1) = 0, \quad$ where $r(x) = 1$ if $x < \frac{1}{2},$
      $r(x) = 4$ if $x > \frac{1}{2}.$

8.   For the problem of approximating the periodic solution of $\ddot{x} + x^3 = \sin t$, discussed in Section 5.7, obtain the second Galerkin approximation

$$x_2(t) = (1.434 \cdots)\sin t + (-0.124 \cdots)\sin 3t.$$

(*Hint:* Obtain the Galerkin equations $f_i(b_1, b_3) = 0, \quad i = 1, 2,$ then use the previous value of b_1 to obtain a value of b_3, then use a numerical technique such as iteration or Newton's method to obtain the corrected values—have patience.)

9.   For the forced Van der Pol equation

$$\ddot{x} - 0.1(1 - x^2)\dot{x} + x = 0.1 \sin t,$$

obtain a first Galerkin approximation

$$x_1(t) = (-0.142 \cdots)\sin t + (-2.379 \cdots)\cos t.$$

10. For the nonlinear boundary value problem

$$y'' + y + \tfrac{1}{2}y^3 = \frac{x}{2}; \quad y(0) = 0, \quad y'(1) + y(1) = 0,$$

one can compute Galerkin approximations as follows: Let $\{\phi_n(x)\}$ be the normalized eigenfunctions of the equation $y'' + \lambda y = 0$, with the boundary conditions as above. Then assume an approximate solution of the form

$$y_n(x) = \sum_{1}^{n} a_k \phi_k(x),$$

where the a_k are determined by the Galerkin equations. For the problem above obtain a first Galerkin approximation

$$y_1(x) = a_1\phi_1(x).$$

References

1. N. I. Akhieser, *The Calculus of Variations*, translated by A. H. Frink, Blaisdell, New York, 1962.
2. G. F. Carrier and C. E. Pearson, *Ordinary Differential Equations*, Blaisdell, Waltham, Mass., 1968.
3. L. Cesari,

    Functional analysis and periodic solutions of nonlinear differential equations, Contributions to Differential Equations I, Wiley, New York, 1963, pp. 149–187.

    Functional analysis and Galerkin's method, Michigan Mathematics Journal *11* (1964), 385–414.

    Functional analysis and differential equations, Studies in Applied Mathematics 5, Soc. Ind. and Appl. Math. (1969), 143–155.

4. L. Collatz, *The Numerical Treatment of Differential Equations*, 3rd Edition, translated by P. G. Williams, Springer-Verlag, Berlin, 1966.
5. R. Courant and D. Hilbert, *Methods of Mathematical Physics*, Vol. 1, Interscience, New York, 1953.
6. I. M. Gelfand and S. V. Fomin, *Calculus of Variations*, translated by R. A. Silverman, Prentice-Hall, Englewood Cliffs, N. J., 1963.
7. F. B. Hildebrand, *Methods of Applied Mathematics*, 2nd Edition, Prentice-Hall, Englewood Cliffs, N. J., 1965.
8. L. V. Kantorovich and V. I. Krylov, *Approximate Methods of Higher Analysis*, translated by C. D. Benster, Interscience, New York, 1958.
9. H. B. Keller, *Numerical Methods for Two-Point Boundary Value Problems*, Blaisdell, Waltham, Mass., 1968.
10. W. Leighton, *Ordinary Differential Equations*, 2nd Edition, Wadsworth, Belmont, California, 1966.

11.   H. Sagan, *Boundary and Eigenvalue Problems in Mathematical Physics*, Wiley, New York, 1961.

12.   M. Urabe and A. Reiter, *Numerical computation of nonlinear forced oscillations by Galerkin's procedure*, Journal of Mathematical Analysis and Application *14* (1966), 107–140.

13.   K. Yosida, *Lectures on Differential and Integral Equations*, Interscience, New York, 1960.

Index

151

A CATALOGUE OF
SELECTED DOVER BOOKS
IN ALL FIELDS OF INTEREST

A CATALOGUE OF SELECTED DOVER
BOOKS IN ALL FIELDS OF INTEREST

RACKHAM'S COLOR ILLUSTRATIONS FOR WAGNER'S RING. Rackham's finest mature work—all 64 full-color watercolors in a faithful and lush interpretation of the *Ring*. Full-sized plates on coated stock of the paintings used by opera companies for authentic staging of Wagner. Captions aid in following complete Ring cycle. Introduction. 64 illustrations plus vignettes. 72pp. 8⅝ x 11¼.　　　　　　　　23779-6 Pa. $6.00

CONTEMPORARY POLISH POSTERS IN FULL COLOR, edited by Joseph Czestochowski. 46 full-color examples of brilliant school of Polish graphic design, selected from world's first museum (near Warsaw) dedicated to poster art. Posters on circuses, films, plays, concerts all show cosmopolitan influences, free imagination. Introduction. 48pp. 9⅜ x 12¼.
23780-X Pa. $6.00

GRAPHIC WORKS OF EDVARD MUNCH, Edvard Munch. 90 haunting, evocative prints by first major Expressionist artist and one of the greatest graphic artists of his time: *The Scream, Anxiety, Death Chamber, The Kiss, Madonna,* etc. Introduction by Alfred Werner. 90pp. 9 x 12.
23765-6 Pa. $5.00

THE GOLDEN AGE OF THE POSTER, Hayward and Blanche Cirker. 70 extraordinary posters in full colors, from Maitres de l'Affiche, Mucha, Lautrec, Bradley, Cheret, Beardsley, many others. Total of 78pp. 9⅜ x 12¼.　　　　　　　　　　　　22753-7 Pa. $5.95

THE NOTEBOOKS OF LEONARDO DA VINCI, edited by J. P. Richter. Extracts from manuscripts reveal great genius; on painting, sculpture, anatomy, sciences, geography, etc. Both Italian and English. 186 ms. pages reproduced, plus 500 additional drawings, including studies for *Last Supper,* Sforza monument, etc. 860pp. 7⅞ x 10¾. (Available in U.S. only)
22572-0, 22573-9 Pa., Two-vol. set $15.90

THE CODEX NUTTALL, as first edited by Zelia Nuttall. Only inexpensive edition, in full color, of a pre-Columbian Mexican (Mixtec) book. 88 color plates show kings, gods, heroes, temples, sacrifices. New explanatory, historical introduction by Arthur G. Miller. 96pp. 11⅜ x 8½. (Available in U.S. only)　　　　　　　　　　23168-2 Pa. $7.95

UNE SEMAINE DE BONTÉ, A SURREALISTIC NOVEL IN COLLAGE, Max Ernst. Masterpiece created out of 19th-century periodical illustrations, explores worlds of terror and surprise. Some consider this Ernst's greatest work. 208pp. 8⅛ x 11.　　　　　　　　23252-2 Pa. $6.00

DRAWINGS OF WILLIAM BLAKE, William Blake. 92 plates from Book of Job, *Divine Comedy, Paradise Lost,* visionary heads, mythological figures, Laocoon, etc. Selection, introduction, commentary by Sir Geoffrey Keynes. 178pp. 8⅛ x 11. 22303-5 Pa. $4.00

ENGRAVINGS OF HOGARTH, William Hogarth. 101 of Hogarth's greatest works: *Rake's Progress, Harlot's Progress, Illustrations for Hudibras, Before and After, Beer Street and Gin Lane,* many more. Full commentary. 256pp. 11 x 13¾. 22479-1 Pa. $12.95

DAUMIER: 120 GREAT LITHOGRAPHS, Honore Daumier. Wide-ranging collection of lithographs by the greatest caricaturist of the 19th century. Concentrates on eternally popular series on lawyers, on married life, on liberated women, etc. Selection, introduction, and notes on plates by Charles F. Ramus. Total of 158pp. 9⅜ x 12¼. 23512-2 Pa. $6.00

DRAWINGS OF MUCHA, Alphonse Maria Mucha. Work reveals draftsman of highest caliber: studies for famous posters and paintings, renderings for book illustrations and ads, etc. 70 works, 9 in color; including 6 items not drawings. Introduction. List of illustrations. 72pp. 9⅜ x 12¼. (Available in U.S. only) 23672-2 Pa. $4.00

GIOVANNI BATTISTA PIRANESI: DRAWINGS IN THE PIERPONT MORGAN LIBRARY, Giovanni Battista Piranesi. For first time ever all of Morgan Library's collection, world's largest. 167 illustrations of rare Piranesi drawings—archeological, architectural, decorative and visionary. Essay, detailed list of drawings, chronology, captions. Edited by Felice Stampfle. 144pp. 9⅜ x 12¼. 23714-1 Pa. $7.50

NEW YORK ETCHINGS (1905-1949), John Sloan. All of important American artist's N.Y. life etchings. 67 works include some of his best art; also lively historical record—Greenwich Village, tenement scenes. Edited by Sloan's widow. Introduction and captions. 79pp. 8⅜ x 11¼. 23651-X Pa. $4.00

CHINESE PAINTING AND CALLIGRAPHY: A PICTORIAL SURVEY, Wan-go Weng. 69 fine examples from John M. Crawford's matchless private collection: landscapes, birds, flowers, human figures, etc., plus calligraphy. Every basic form included: hanging scrolls, handscrolls, album leaves, fans, etc. 109 illustrations. Introduction. Captions. 192pp. 8⅞ x 11¾. 23707-9 Pa. $7.95

DRAWINGS OF REMBRANDT, edited by Seymour Slive. Updated Lippmann, Hofstede de Groot edition, with definitive scholarly apparatus. All portraits, biblical sketches, landscapes, nudes, Oriental figures, classical studies, together with selection of work by followers. 550 illustrations. Total of 630pp. 9⅛ x 12¼. 21485-0, 21486-9 Pa., Two-vol. set $15.00

THE DISASTERS OF WAR, Francisco Goya. 83 etchings record horrors of Napoleonic wars in Spain and war in general. Reprint of 1st edition, plus 3 additional plates. Introduction by Philip Hofer. 97pp. 9⅜ x 8¼. 21872-4 Pa. $4.00

THE EARLY WORK OF AUBREY BEARDSLEY, Aubrey Beardsley. 157 plates, 2 in color: *Manon Lescaut, Madame Bovary, Morte Darthur, Salome,* other. Introduction by H. Marillier. 182pp. 8⅛ x 11. 21816-3 Pa. $4.50

THE LATER WORK OF AUBREY BEARDSLEY, Aubrey Beardsley. Exotic masterpieces of full maturity: *Venus and Tannhauser, Lysistrata, Rape of the Lock, Volpone,* Savoy material, etc. 174 plates, 2 in color. 186pp. 8⅛ x 11. 21817-1 Pa. $5.95

THOMAS NAST'S CHRISTMAS DRAWINGS, Thomas Nast. Almost all Christmas drawings by creator of image of Santa Claus as we know it, and one of America's foremost illustrators and political cartoonists. 66 illustrations. 3 illustrations in color on covers. 96pp. 8⅜ x 11¼. 23660-9 Pa. $3.50

THE DORÉ ILLUSTRATIONS FOR DANTE'S DIVINE COMEDY, Gustave Doré. All 135 plates from Inferno, Purgatory, Paradise; fantastic tortures, infernal landscapes, celestial wonders. Each plate with appropriate (translated) verses. 141pp. 9 x 12. 23231-X Pa. $4.50

DORÉ'S ILLUSTRATIONS FOR RABELAIS, Gustave Doré. 252 striking illustrations of *Gargantua and Pantagruel* books by foremost 19th-century illustrator. Including 60 plates, 192 delightful smaller illustrations. 153pp. 9 x 12. 23656-0 Pa. $5.00

LONDON: A PILGRIMAGE, Gustave Doré, Blanchard Jerrold. Squalor, riches, misery, beauty of mid-Victorian metropolis; 55 wonderful plates, 125 other illustrations, full social, cultural text by Jerrold. 191pp. of text. 9⅜ x 12¼. 22306-X Pa. $7.00

THE RIME OF THE ANCIENT MARINER, Gustave Doré, S. T. Coleridge. Dore's finest work, 34 plates capture moods, subtleties of poem. Full text. Introduction by Millicent Rose. 77pp. 9¼ x 12. 22305-1 Pa. $3.50

THE DORE BIBLE ILLUSTRATIONS, Gustave Doré. All wonderful, detailed plates: Adam and Eve, Flood, Babylon, Life of Jesus, etc. Brief King James text with each plate. Introduction by Millicent Rose. 241 plates. 241pp. 9 x 12. 23004-X Pa. $6.00

THE COMPLETE ENGRAVINGS, ETCHINGS AND DRYPOINTS OF ALBRECHT DURER. "Knight, Death and Devil"; "Melencolia," and more—all Dürer's known works in all three media, including 6 works formerly attributed to him. 120 plates. 235pp. 8⅜ x 11¼. 22851-7 Pa. $6.50

MECHANICK EXERCISES ON THE WHOLE ART OF PRINTING, Joseph Moxon. First complete book (1683-4) ever written about typography, a compendium of everything known about printing at the latter part of 17th century. Reprint of 2nd (1962) Oxford Univ. Press edition. 74 illustrations. Total of 550pp. 6⅛ x 9¼. 23617-X Pa. $7.95

THE COMPLETE WOODCUTS OF ALBRECHT DURER, edited by Dr. W. Kurth. 346 in all: "Old Testament," "St. Jerome," "Passion," "Life of Virgin," Apocalypse," many others. Introduction by Campbell Dodgson. 285pp. 8½ x 12¼. 21097-9 Pa. $7.50

DRAWINGS OF ALBRECHT DURER, edited by Heinrich Wolfflin. 81 plates show development from youth to full style. Many favorites; many new. Introduction by Alfred Werner. 96pp. 8⅛ x 11. 22352-3 Pa. $5.00

THE HUMAN FIGURE, Albrecht Dürer. Experiments in various techniques—stereometric, progressive proportional, and others. Also life studies that rank among finest ever done. Complete reprinting of *Dresden Sketchbook*. 170 plates. 355pp. 8⅜ x 11¼. 21042-1 Pa. $7.95

OF THE JUST SHAPING OF LETTERS, Albrecht Dürer. Renaissance artist explains design of Roman majuscules by geometry, also Gothic lower and capitals. Grolier Club edition. 43pp. 7⅞ x 10¾ 21306-4 Pa. $3.00

TEN BOOKS ON ARCHITECTURE, Vitruvius. The most important book ever written on architecture. Early Roman aesthetics, technology, classical orders, site selection, all other aspects. Stands behind everything since. Morgan translation. 331pp. 5⅜ x 8½. 20645-9 Pa. $4.50

THE FOUR BOOKS OF ARCHITECTURE, Andrea Palladio. 16th-century classic responsible for Palladian movement and style. Covers classical architectural remains, Renaissance revivals, classical orders, etc. 1738 Ware English edition. Introduction by A. Placzek. 216 plates. 110pp. of text. 9½ x 12¾. 21308-0 Pa. $10.00

HORIZONS, Norman Bel Geddes. Great industrialist stage designer, "father of streamlining," on application of aesthetics to transportation, amusement, architecture, etc. 1932 prophetic account; function, theory, specific projects. 222 illustrations. 312pp. 7⅞ x 10¾. 23514-9 Pa. $6.95

FRANK LLOYD WRIGHT'S FALLINGWATER, Donald Hoffmann. Full, illustrated story of conception and building of Wright's masterwork at Bear Run, Pa. 100 photographs of site, construction, and details of completed structure. 112pp. 9¼ x 10. 23671-4 Pa. $5.50

THE ELEMENTS OF DRAWING, John Ruskin. Timeless classic by great Viltorian; starts with basic ideas, works through more difficult. Many practical exercises. 48 illustrations. Introduction by Lawrence Campbell. 228pp. 5⅜ x 8½. 22730-8 Pa. $3.75

GIST OF ART, John Sloan. Greatest modern American teacher, Art Students League, offers innumerable hints, instructions, guided comments to help you in painting. Not a formal course. 46 illustrations. Introduction by Helen Sloan. 200pp. 5⅜ x 8½. 23435-5 Pa. $4.00

THE ANATOMY OF THE HORSE, George Stubbs. Often considered the great masterpiece of animal anatomy. Full reproduction of 1766 edition, plus prospectus; original text and modernized text. 36 plates. Introduction by Eleanor Garvey. 121pp. 11 x 14¾. 23402-9 Pa. $6.00

BRIDGMAN'S LIFE DRAWING, George B. Bridgman. More than 500 illustrative drawings and text teach you to abstract the body into its major masses, use light and shade, proportion; as well as specific areas of anatomy, of which Bridgman is master. 192pp. 6½ x 9¼. (Available in U.S. only) 22710-3 Pa. $3.50

ART NOUVEAU DESIGNS IN COLOR, Alphonse Mucha, Maurice Verneuil, Georges Auriol. Full-color reproduction of *Combinaisons ornementales* (c. 1900) by Art Nouveau masters. Floral, animal, geometric, interlacings, swashes—borders, frames, spots—all incredibly beautiful. 60 plates, hundreds of designs. 9⅜ x 8-1/16. 22885-1 Pa. $4.00

FULL-COLOR FLORAL DESIGNS IN THE ART NOUVEAU STYLE, E. A. Seguy. 166 motifs, on 40 plates, from *Les fleurs et leurs applications decoratives* (1902): borders, circular designs, repeats, allovers, "spots." All in authentic Art Nouveau colors. 48pp. 9⅜ x 12¼. 23439-8 Pa. $5.00

A DIDEROT PICTORIAL ENCYCLOPEDIA OF TRADES AND IN-DUSTRY, edited by Charles C. Gillispie. 485 most interesting plates from the great French Encyclopedia of the 18th century show hundreds of working figures, artifacts, process, land and cityscapes; glassmaking, paper-making, metal extraction, construction, weaving, making furniture, clothing, wigs, dozens of other activities. Plates fully explained. 920pp. 9 x 12. 22284-5, 22285-3 Clothbd., Two-vol. set $40.00

HANDBOOK OF EARLY ADVERTISING ART, Clarence P. Hornung. Largest collection of copyright-free early and antique advertising art ever compiled. Over 6,000 illustrations, from Franklin's time to the 1890's for special effects, novelty. Valuable source, almost inexhaustible.
Pictorial Volume. Agriculture, the zodiac, animals, autos, birds, Christmas, fire engines, flowers, trees, musical instruments, ships, games and sports, much more. Arranged by subject matter and use. 237 plates. 288pp. 9 x 12. 20122-8 Clothbd. $14.50

Typographical Volume. Roman and Gothic faces ranging from 10 point to 300 point, "Barnum," German and Old English faces, script, logotypes, scrolls and flourishes, 1115 ornamental initials, 67 complete alphabets, more. 310 plates. 320pp. 9 x 12. 20123-6 Clothbd. $15.00

CALLIGRAPHY (CALLIGRAPHIA LATINA), J. G. Schwandner. High point of 18th-century ornamental calligraphy. Very ornate initials, scrolls, borders, cherubs, birds, lettered examples. 172pp. 9 x 13. 20475-8 Pa. $7.00

ART FORMS IN NATURE, Ernst Haeckel. Multitude of strangely beautiful natural forms: Radiolaria, Foraminifera, jellyfishes, fungi, turtles, bats, etc. All 100 plates of the 19th-century evolutionist's *Kunstformen der Natur* (1904). 100pp. 9⅜ x 12¼. 22987-4 Pa. $5.00

CHILDREN: A PICTORIAL ARCHIVE FROM NINETEENTH-CENTURY SOURCES, edited by Carol Belanger Grafton. 242 rare, copyright-free wood engravings for artists and designers. Widest such selection available. All illustrations in line. 119pp. 8⅜ x 11¼.
23694-3 Pa. $4.00

WOMEN: A PICTORIAL ARCHIVE FROM NINETEENTH-CENTURY SOURCES, edited by Jim Harter. 391 copyright-free wood engravings for artists and designers selected from rare periodicals. Most extensive such collection available. All illustrations in line. 128pp. 9 x 12.
23703-6 Pa. $4.50

ARABIC ART IN COLOR, Prisse d'Avennes. From the greatest ornamentalists of all time—50 plates in color, rarely seen outside the Near East, rich in suggestion and stimulus. Includes 4 plates on covers. 46pp. 9⅜ x 12¼. 23658-7 Pa. $6.00

AUTHENTIC ALGERIAN CARPET DESIGNS AND MOTIFS, edited by June Beveridge. Algerian carpets are world famous. Dozens of geometrical motifs are charted on grids, color-coded, for weavers, needleworkers, craftsmen, designers. 53 illustrations plus 4 in color. 48pp. 8¼ x 11. (Available in U.S. only) 23650-1 Pa. $1.75

DICTIONARY OF AMERICAN PORTRAITS, edited by Hayward and Blanche Cirker. 4000 important Americans, earliest times to 1905, mostly in clear line. Politicians, writers, soldiers, scientists, inventors, industrialists, Indians, Blacks, women, outlaws, etc. Identificatory information. 756pp. 9¼ x 12¾. 21823-6 Clothbd. $40.00

HOW THE OTHER HALF LIVES, Jacob A. Riis. Journalistic record of filth, degradation, upward drive in New York immigrant slums, shops, around 1900. New edition includes 100 original Riis photos, monuments of early photography. 233pp. 10 x 7⅞. 22012-5 Pa. $7.00

NEW YORK IN THE THIRTIES, Berenice Abbott. Noted photographer's fascinating study of city shows new buildings that have become famous and old sights that have disappeared forever. Insightful commentary. 97 photographs. 97pp. 11⅜ x 10. 22967-X Pa. $5.00

MEN AT WORK, Lewis W. Hine. Famous photographic studies of construction workers, railroad men, factory workers and coal miners. New supplement of 18 photos on Empire State building construction. New introduction by Jonathan L. Doherty. Total of 69 photos. 63pp. 8 x 10¾.
23475-4 Pa. $3.00

THE DEPRESSION YEARS AS PHOTOGRAPHED BY ARTHUR ROTH-STEIN, Arthur Rothstein. First collection devoted entirely to the work of outstanding 1930s photographer: famous dust storm photo, ragged children, unemployed, etc. 120 photographs. Captions. 119pp. 9¼ x 10¾.
23590-4 Pa. $5.00

CAMERA WORK: A PICTORIAL GUIDE, Alfred Stieglitz. All 559 illustrations and plates from the most important periodical in the history of art photography, Camera Work (1903-17). Presented four to a page, reduced in size but still clear, in strict chronological order, with complete captions. Three indexes. Glossary. Bibliography. 176pp. 8⅜ x 11¼.
23591-2 Pa. $6.95

ALVIN LANGDON COBURN, PHOTOGRAPHER, Alvin L. Coburn. Revealing autobiography by one of greatest photographers of 20th century gives insider's version of Photo-Secession, plus comments on his own work. 77 photographs by Coburn. Edited by Helmut and Alison Gernsheim. 160pp. 8⅛ x 11.
23685-4 Pa. $6.00

NEW YORK IN THE FORTIES, Andreas Feininger. 162 brilliant photographs by the well-known photographer, formerly with Life magazine, show commuters, shoppers, Times Square at night, Harlem nightclub, Lower East Side, etc. Introduction and full captions by John von Hartz. 181pp. 9¼ x 10¾.
23585-8 Pa. $6.95

GREAT NEWS PHOTOS AND THE STORIES BEHIND THEM, John Faber. Dramatic volume of 140 great news photos, 1855 through 1976, and revealing stories behind them, with both historical and technical information. Hindenburg disaster, shooting of Oswald, nomination of Jimmy Carter, etc. 160pp. 8¼ x 11.
23667-6 Pa. $5.00

THE ART OF THE CINEMATOGRAPHER, Leonard Maltin. Survey of American cinematography history and anecdotal interviews with 5 masters—Arthur Miller, Hal Mohr, Hal Rosson, Lucien Ballard, and Conrad Hall. Very large selection of behind-the-scenes production photos. 105 photographs. Filmographies. Index. Originally Behind the Camera. 144pp. 8¼ x 11.
23686-2 Pa. $5.00

DESIGNS FOR THE THREE-CORNERED HAT (LE TRICORNE), Pablo Picasso. 32 fabulously rare drawings—including 31 color illustrations of costumes and accessories—for 1919 production of famous ballet. Edited by Parmenia Migel, who has written new introduction. 48pp. 9⅜ x 12¼.
(Available in U.S. only)
23709-5 Pa. $5.00

NOTES OF A FILM DIRECTOR, Sergei Eisenstein. Greatest Russian filmmaker explains montage, making of Alexander Nevsky, aesthetics; comments on self, associates, great rivals (Chaplin), similar material. 78 illustrations. 240pp. 5⅜ x 8½.
22392-2 Pa. $4.50

HOLLYWOOD GLAMOUR PORTRAITS, edited by John Kobal. 145 photos capture the stars from 1926-49, the high point in portrait photography. Gable, Harlow, Bogart, Bacall, Hedy Lamarr, Marlene Dietrich, Robert Montgomery, Marlon Brando, Veronica Lake; 94 stars in all. Full background on photographers, technical aspects, much more. Total of 160pp. 8⅜ x 11¼.   23352-9 Pa. $6.00

THE NEW YORK STAGE: FAMOUS PRODUCTIONS IN PHOTO-GRAPHS, edited by Stanley Appelbaum. 148 photographs from Museum of City of New York show 142 plays, 1883-1939. *Peter Pan, The Front Page, Dead End, Our Town,* O'Neill, hundreds of actors and actresses, etc. Full indexes. 154pp. 9½ x 10.   23241-7 Pa. $6.00

DIALOGUES CONCERNING TWO NEW SCIENCES, Galileo Galilei. Encompassing 30 years of experiment and thought, these dialogues deal with geometric demonstrations of fracture of solid bodies, cohesion, lever-age, speed of light and sound, pendulums, falling bodies, accelerated motion, etc. 300pp. 5⅜ x 8½.   60099-8 Pa. $4.00

THE GREAT OPERA STARS IN HISTORIC PHOTOGRAPHS, edited by James Camner. 343 portraits from the 1850s to the 1940s: Tamburini, Mario, Caliapin, Jeritza, Melchior, Melba, Patti, Pinza, Schipa, Caruso, Farrar, Steber, Gobbi, and many more—270 performers in all. Index. 199pp. 8⅜ x 11¼.   23575-0 Pa. $7.50

J. S. BACH, Albert Schweitzer. Great full-length study of Bach, life, background to music, music, by foremost modern scholar. Ernest Newman translation. 650 musical examples. Total of 928pp. 5⅜ x 8½. (Available in U.S. only)   21631-4, 21632-2 Pa., Two-vol. set $11.00

COMPLETE PIANO SONATAS, Ludwig van Beethoven. All sonatas in the fine Schenker edition, with fingering, analytical material. One of best modern editions. Total of 615pp. 9 x 12. (Available in U.S. only)
                              23134-8, 23135-6 Pa., Two-vol. set $15.50

KEYBOARD MUSIC, J. S. Bach. Bach-Gesellschaft edition. For harpsi-chord, piano, other keyboard instruments. English Suites, French Suites, Six Partitas, Goldberg Variations, Two-Part Inventions, Three-Part Sin-fonias. 312pp. 8⅛ x 11. (Available in U.S. only)   22360-4 Pa. $6.95

FOUR SYMPHONIES IN FULL SCORE, Franz Schubert. Schubert's four most popular symphonies: No. 4 in C Minor ("Tragic"); No. 5 in B-flat Major; No. 8 in B Minor ("Unfinished"); No. 9 in C Major ("Great"). Breitkopf & Hartel edition. Study score. 261pp. 9⅜ x 12¼.
                                                        23681-1 Pa. $6.50

THE AUTHENTIC GILBERT & SULLIVAN SONGBOOK, W. S. Gilbert, A. S. Sullivan. Largest selection available; 92 songs, uncut, original keys, in piano rendering approved by Sullivan. Favorites and lesser-known fine numbers. Edited with plot synopses by James Spero. 3 illustrations. 399pp. 9 x 12.   23482-7 Pa. $9.95

PRINCIPLES OF ORCHESTRATION, Nikolay Rimsky-Korsakov. Great classical orchestrator provides fundamentals of tonal resonance, progression of parts, voice and orchestra, tutti effects, much else in major document. 330pp. of musical excerpts. 489pp. 6½ x 9¼. 21266-1 Pa. $7.50

TRISTAN UND ISOLDE, Richard Wagner. Full orchestral score with complete instrumentation. Do not confuse with piano reduction. Commentary by Felix Mottl, great Wagnerian conductor and scholar. Study score. 655pp. 8⅛ x 11. 22915-7 Pa. $13.95

REQUIEM IN FULL SCORE, Giuseppe Verdi. Immensely popular with choral groups and music lovers. Republication of edition published by C. F. Peters, Leipzig, n. d. German frontmaker in English translation. Glossary. Text in Latin. Study score. 204pp. 9⅜ x 12¼.
23682-X Pa. $6.00

COMPLETE CHAMBER MUSIC FOR STRINGS, Felix Mendelssohn. All of Mendelssohn's chamber music: Octet, 2 Quintets, 6 Quartets, and Four Pieces for String Quartet. (Nothing with piano is included). Complete works edition (1874-7). Study score. 283 pp. 9⅜ x 12¼.
23679-X Pa. $7.50

POPULAR SONGS OF NINETEENTH-CENTURY AMERICA, edited by Richard Jackson. 64 most important songs: "Old Oaken Bucket," "Arkansas Traveler," "Yellow Rose of Texas," etc. Authentic original sheet music, full introduction and commentaries. 290pp. 9 x 12. 23270-0 Pa. $7.95

COLLECTED PIANO WORKS, Scott Joplin. Edited by Vera Brodsky Lawrence. Practically all of Joplin's piano works—rags, two-steps, marches, waltzes, etc., 51 works in all. Extensive introduction by Rudi Blesh. Total of 345pp. 9 x 12. 23106-2 Pa. $14.95

BASIC PRINCIPLES OF CLASSICAL BALLET, Agrippina Vaganova. Great Russian theoretician, teacher explains methods for teaching classical ballet; incorporates best from French, Italian, Russian schools. 118 illustrations. 175pp. 5⅜ x 8½. 22036-2 Pa. $2.50

CHINESE CHARACTERS, L. Wieger. Rich analysis of 2300 characters according to traditional systems into primitives. Historical-semantic analysis to phonetics (Classical Mandarin) and radicals. 820pp. 6⅛ x 9¼.
21321-8 Pa. $10.00

EGYPTIAN LANGUAGE: EASY LESSONS IN EGYPTIAN HIERO-GLYPHICS, E. A. Wallis Budge. Foremost Egyptologist offers Egyptian grammar, explanation of hieroglyphics, many reading texts, dictionary of symbols. 246pp. 5 x 7½. (Available in U.S. only)
21394-3 Clothbd. $7.50

AN ETYMOLOGICAL DICTIONARY OF MODERN ENGLISH, Ernest Weekley. Richest, fullest work, by foremost British lexicographer. Detailed word histories. Inexhaustible. Do not confuse this with Concise Etymological Dictionary, which is abridged. Total of 856pp. 6½ x 9¼.
21873-2, 21874-0 Pa., Two-vol. set $12.00

A MAYA GRAMMAR, Alfred M. Tozzer. Practical, useful English-language grammar by the Harvard anthropologist who was one of the three greatest American scholars in the area of Maya culture. Phonetics, grammatical processes, syntax, more. 301pp. 5⅜ x 8½.   23465-7 Pa. $4.00

THE JOURNAL OF HENRY D. THOREAU, edited by Bradford Torrey, F. H. Allen. Complete reprinting of 14 volumes, 1837-61, over two million words; the sourcebooks for *Walden,* etc. Definitive. All original sketches, plus 75 photographs. Introduction by Walter Harding. Total of 1804pp. 8½ x 12¼.   20312-3, 20313-1 Clothbd., Two-vol. set $70.00

CLASSIC GHOST STORIES, Charles Dickens and others. 18 wonderful stories you've wanted to reread: "The Monkey's Paw," "The House and the Brain," "The Upper Berth," "The Signalman," "Dracula's Guest," "The Tapestried Chamber," etc. Dickens, Scott, Mary Shelley, Stoker, etc. 330pp. 5⅜ x 8½.   20735-8 Pa. $4.50

SEVEN SCIENCE FICTION NOVELS, H. G. Wells. Full novels. *First Men in the Moon, Island of Dr. Moreau, War of the Worlds, Food of the Gods, Invisible Man, Time Machine, In the Days of the Comet.* A basic science-fiction library. 1015pp. 5⅜ x 8½. (Available in U.S. only)
20264-X Clothbd. $8.95

ARMADALE, Wilkie Collins. Third great mystery novel by the author of *The Woman in White* and *The Moonstone.* Ingeniously plotted narrative shows an exceptional command of character, incident and mood. Original magazine version with 40 illustrations. 597pp. 5⅜ x 8½.
23429-0 Pa. $6.00

MASTERS OF MYSTERY, H. Douglas Thomson. The first book in English (1931) devoted to history and aesthetics of detective story. Poe, Doyle, LeFanu, Dickens, many others, up to 1930. New introduction and notes by E. F. Bleiler. 288pp. 5⅜ x 8½. (Available in U.S. only)
23606-4 Pa. $4.00

FLATLAND, E. A. Abbott. Science-fiction classic explores life of 2-D being in 3-D world. Read also as introduction to thought about hyperspace. Introduction by Banesh Hoffmann. 16 illustrations. 103pp. 5⅜ x 8½.
20001-9 Pa. $2.00

THREE SUPERNATURAL NOVELS OF THE VICTORIAN PERIOD, edited, with an introduction, by E. F. Bleiler. Reprinted complete and unabridged, three great classics of the supernatural: *The Haunted Hotel* by Wilkie Collins, *The Haunted House at Latchford* by Mrs. J. H. Riddell, and *The Lost Stradivarious* by J. Meade Falkner. 325pp. 5⅜ x 8½.
22571-2 Pa. $4.00

AYESHA: THE RETURN OF "SHE," H. Rider Haggard. Virtuoso sequel featuring the great mythic creation, Ayesha, in an adventure that is fully as good as the first book, *She.* Original magazine version, with 47 original illustrations by Maurice Greiffenhagen. 189pp. 6½ x 9¼.
23649-8 Pa. $3.50

UNCLE SILAS, J. Sheridan LeFanu. Victorian Gothic mystery novel, considered by many best of period, even better than Collins or Dickens. Wonderful psychological terror. Introduction by Frederick Shroyer. 436pp. 5⅜ x 8½.   21715-9 Pa. $6.00

JURGEN, James Branch Cabell. The great erotic fantasy of the 1920's that delighted thousands, shocked thousands more. Full final text, Lane edition with 13 plates by Frank Pape. 346pp. 5⅜ x 8½.
23507-6 Pa. $4.50

THE CLAVERINGS, Anthony Trollope. Major novel, chronicling aspects of British Victorian society, personalities. Reprint of Cornhill serialization, 16 plates by M. Edwards; first reprint of full text. Introduction by Norman Donaldson. 412pp. 5⅜ x 8½.   23464-9 Pa. $5.00

KEPT IN THE DARK, Anthony Trollope. Unusual short novel about Victorian morality and abnormal psychology by the great English author. Probably the first American publication. Frontispiece by Sir John Millais. 92pp. 6½ x 9¼.   23609-9 Pa. $2.50

RALPH THE HEIR, Anthony Trollope. Forgotten tale of illegitimacy, inheritance. Master novel of Trollope's later years. Victorian country estates, clubs, Parliament, fox hunting, world of fully realized characters. Reprint of 1871 edition. 12 illustrations by F. A. Faser. 434pp. of text. 5⅜ x 8½.   23642-0 Pa. $5.00

YEKL and THE IMPORTED BRIDEGROOM AND OTHER STORIES OF THE NEW YORK GHETTO, Abraham Cahan. Film *Hester Street* based on *Yekl* (1896). Novel, other stories among first about Jewish immigrants of N.Y.'s East Side. Highly praised by W. D. Howells—Cahan "a new star of realism." New introduction by Bernard G. Richards. 240pp. 5⅜ x 8½.   22427-9 Pa. $3.50

THE HIGH PLACE, James Branch Cabell. Great fantasy writer's enchanting comedy of disenchantment set in 18th-century France. Considered by some critics to be even better than his famous *Jurgen*. 10 illustrations and numerous vignettes by noted fantasy artist Frank C. Pape. 320pp. 5⅜ x 8½.   23670-6 Pa. $4.00

ALICE'S ADVENTURES UNDER GROUND, Lewis Carroll. Facsimile of ms. Carroll gave Alice Liddell in 1864. Different in many ways from final Alice. Handlettered, illustrated by Carroll. Introduction by Martin Gardner. 128pp. 5⅜ x 8½.   21482-6 Pa. $2.50

FAVORITE ANDREW LANG FAIRY TALE BOOKS IN MANY COLORS, Andrew Lang. The four Lang favorites in a boxed set—the complete *Red, Green, Yellow* and *Blue* Fairy Books. 164 stories; 439 illustrations by Lancelot Speed, Henry Ford and G. P. Jacomb Hood. Total of about 1500pp. 5⅜ x 8½.   23407-X Boxed set, Pa. $15.95

HOUSEHOLD STORIES BY THE BROTHERS GRIMM. All the great Grimm stories: "Rumpelstiltskin," "Snow White," "Hansel and Gretel," etc., with 114 illustrations by Walter Crane. 269pp. 5⅜ x 8½.
21080-4 Pa. $3.50

SLEEPING BEAUTY, illustrated by Arthur Rackham. Perhaps the fullest, most delightful version ever, told by C. S. Evans. Rackham's best work. 49 illustrations. 110pp. 7⅞ x 10¾.
22756-1 Pa. $2.50

AMERICAN FAIRY TALES, L. Frank Baum. Young cowboy lassoes Father Time; dummy in Mr. Floman's department store window comes to life; and 10 other fairy tales. 41 illustrations by N. P. Hall, Harry Kennedy, Ike Morgan, and Ralph Gardner. 209pp. 5⅜ x 8½.
23643-9 Pa. $3.00

THE WONDERFUL WIZARD OF OZ, L. Frank Baum. Facsimile in full color of America's finest children's classic. Introduction by Martin Gardner. 143 illustrations by W. W. Denslow. 267pp. 5⅜ x 8½.
20691-2 Pa. $3.50

THE TALE OF PETER RABBIT, Beatrix Potter. The inimitable Peter's terrifying adventure in Mr. McGregor's garden, with all 27 wonderful, full-color Potter illustrations. 55pp. 4¼ x 5½. (Available in U.S. only)
22827-4 Pa. $1.25

THE STORY OF KING ARTHUR AND HIS KNIGHTS, Howard Pyle. Finest children's version of life of King Arthur. 48 illustrations by Pyle. 131pp. 6⅛ x 9¼.
21445-1 Pa. $4.95

CARUSO'S CARICATURES, Enrico Caruso. Great tenor's remarkable caricatures of self, fellow musicians, composers, others. Toscanini, Puccini, Farrar, etc. Impish, cutting, insightful. 473 illustrations. Preface by M. Sisca. 217pp. 8⅜ x 11¼.
23528-9 Pa. $6.95

PERSONAL NARRATIVE OF A PILGRIMAGE TO ALMADINAH AND MECCAH, Richard Burton. Great travel classic by remarkably colorful personality. Burton, disguised as a Moroccan, visited sacred shrines of Islam, narrowly escaping death. Wonderful observations of Islamic life, customs, personalities. 47 illustrations. Total of 959pp. 5⅜ x 8½.
21217-3, 21218-1 Pa., Two-vol. set $12.00

INCIDENTS OF TRAVEL IN YUCATAN, John L. Stephens. Classic (1843) exploration of jungles of Yucatan, looking for evidences of Maya civilization. Travel adventures, Mexican and Indian culture, etc. Total of 669pp. 5⅜ x 8½.
20926-1, 20927-X Pa., Two-vol. set $7.90

AMERICAN LITERARY AUTOGRAPHS FROM WASHINGTON IRVING TO HENRY JAMES, Herbert Cahoon, et al. Letters, poems, manuscripts of Hawthorne, Thoreau, Twain, Alcott, Whitman, 67 other prominent American authors. Reproductions, full transcripts and commentary. Plus checklist of all American Literary Autographs in The Pierpont Morgan Library. Printed on exceptionally high-quality paper. 136 illustrations. 212pp. 9⅛ x 12¼.
23548-3 Pa. $12.50

AN AUTOBIOGRAPHY, Margaret Sanger. Exciting personal account of hard-fought battle for woman's right to birth control, against prejudice, church, law. Foremost feminist document. 504pp. 5⅜ x 8½.
20470-7 Pa. $5.50

MY BONDAGE AND MY FREEDOM, Frederick Douglass. Born as a slave, Douglass became outspoken force in antislavery movement. The best of Douglass's autobiographies. Graphic description of slave life. Introduction by P. Foner. 464pp. 5⅜ x 8½. 22457-0 Pa. $5.50

LIVING MY LIFE, Emma Goldman. Candid, no holds barred account by foremost American anarchist: her own life, anarchist movement, famous contemporaries, ideas and their impact. Struggles and confrontations in America, plus deportation to U.S.S.R. Shocking inside account of persecution of anarchists under Lenin. 13 plates. Total of 944pp. 5⅜ x 8½.
22543-7, 22544-5 Pa., Two-vol. set $12.00

LETTERS AND NOTES ON THE MANNERS, CUSTOMS AND CONDITIONS OF THE NORTH AMERICAN INDIANS, George Catlin. Classic account of life among Plains Indians: ceremonies, hunt, warfare, etc. Dover edition reproduces for first time all original paintings. 312 plates. 572pp. of text. 6⅛ x 9¼. 22118-0, 22119-9 Pa.. Two-vol. set $12.00

THE MAYA AND THEIR NEIGHBORS, edited by Clarence L. Hay, others. Synoptic view of Maya civilization in broadest sense, together with Northern, Southern neighbors. Integrates much background, valuable detail not elsewhere. Prepared by greatest scholars: Kroeber, Morley, Thompson, Spinden, Vaillant, many others. Sometimes called Tozzer Memorial Volume. 60 illustrations, linguistic map. 634pp. 5⅜ x 8½.
23510-6 Pa. $10.00

HANDBOOK OF THE INDIANS OF CALIFORNIA, A. L. Kroeber. Foremost American anthropologist offers complete ethnographic study of each group. Monumental classic. 459 illustrations, maps. 995pp. 5⅜ x 8½.
23368-5 Pa. $13.00

SHAKTI AND SHAKTA, Arthur Avalon. First book to give clear, cohesive analysis of Shakta doctrine, Shakta ritual and Kundalini Shakti (yoga). Important work by one of world's foremost students of Shaktic and Tantric thought. 732pp. 5⅜ x 8½. (Available in U.S. only)
23645-5 Pa. $7.95

AN INTRODUCTION TO THE STUDY OF THE MAYA HIEROGLYPHS, Syvanus Griswold Morley. Classic study by one of the truly great figures in hieroglyph research. Still the best introduction for the student for reading Maya hieroglyphs. New introduction by J. Eric S. Thompson. 117 illustrations. 284pp. 5⅜ x 8½. 23108-9 Pa. $4.00

A STUDY OF MAYA ART, Herbert J. Spinden. Landmark classic interprets Maya symbolism, estimates styles, covers ceramics, architecture, murals, stone carvings as artforms. Still a basic book in area. New introduction by J. Eric Thompson. Over 750 illustrations. 341pp. 8⅜ x 11¼.
21235-1 Pa. $6.95

GEOMETRY, RELATIVITY AND THE FOURTH DIMENSION, Rudolf Rucker. Exposition of fourth dimension, means of visualization, concepts of relativity as Flatland characters continue adventures. Popular, easily followed yet accurate, profound. 141 illustrations. 133pp. 5⅜ x 8½.
23400-2 Pa. $2.75

THE ORIGIN OF LIFE, A. I. Oparin. Modern classic in biochemistry, the first rigorous examination of possible evolution of life from nitrocarbon compounds. Non-technical, easily followed. Total of 295pp. 5⅜ x 8½.
60213-3 Pa. $4.00

PLANETS, STARS AND GALAXIES, A. E. Fanning. Comprehensive introductory survey: the sun, solar system, stars, galaxies, universe, cosmology; quasars, radio stars, etc. 24pp. of photographs. 189pp. 5⅜ x 8½. (Available in U.S. only)
21680-2 Pa. $3.75

THE THIRTEEN BOOKS OF EUCLID'S ELEMENTS, translated with introduction and commentary by Sir Thomas L. Heath. Definitive edition. Textual and linguistic notes, mathematical analysis, 2500 years of critical commentary. Do not confuse with abridged school editions. Total of 1414pp. 5⅜ x 8½.
60088-2, 60089-0, 60090-4 Pa., Three-vol. set $18.50

Prices subject to change without notice.

Available at your book dealer or write for free catalogue to Dept. GI, Dover Publications, Inc., 180 Varick St., N.Y., N.Y. 10014. Dover publishes more than 175 books each year on science, elementary and advanced mathematics, biology, music, art, literary history, social sciences and other areas.